Fusion of Artificial Intelligence and Machine Learning for Advanced Image Processing, Data Analysis, and Cyber Security

AAP Advances in Artificial Intelligence & Robotics

Fusion of Artificial Intelligence and Machine Learning for Advanced Image Processing, Data Analysis, and Cyber Security

Edited By:

Arun Kumar Rana, PhD

Rashmi Gupta, PhD

Sharad Sharma, PhD

Ahmed A. Elngar, PhD

Sachin Dhawan, PhD

AAP | APPLE ACADEMIC PRESS

First edition published 2025

Apple Academic Press Inc.
1265 Goldenrod Circle, NE,
Palm Bay, FL 32905 USA

760 Laurentian Drive, Unit 19,
Burlington, ON L7N 0A4, CANADA

CRC Press
2385 NW Executive Center Drive,
Suite 320, Boca Raton FL 33431

4 Park Square, Milton Park,
Abingdon, Oxon, OX14 4RN UK

Library and Archives Canada Cataloguing in Publication

CIP data on file with Canada Library and Archives

Library of Congress Cataloging-in-Publication Data

CIP data on file with US Library of Congress

ISBN: 978-1-77491-642-1 (hbk)
ISBN: 978-1-77491-643-8 (pbk)
ISBN: 978-1-00349-357-0 (ebk)

AAP ADVANCES IN ARTIFICIAL INTELLIGENCE & ROBOTICS

The new book series AAP Advances in Artificial Intelligence & Robotics will provide detailed coverage of innovations in artificial life, computational intelligence, evolutionary computing, machine learning, robotics, and applications. The list of topics covers all the application areas of artificial intelligence and robotics such as: computational neuroscience, social intelligence, ambient intelligence, artificial life, virtual worlds and society, cognitive science and systems, computational intelligence, human-centered and human-centric computing, intelligent decision making and support, intelligent network security.

With this innovative era of simulated and artificial intelligence, much research is required in order to advance the field and also to estimate the societal and ethical concerns of the existence robotics and scientific computing. The series also aims that books in this series will be practically relevant, so that the results will be useful for managers in leadership roles related to AI and robotics, researchers, data analysts, project managers, and others. Therefore, both theoretical and managerial implications of the research need to be considered.

The book series will broadly consider the contributions from the following fields:

- Artificial Intelligence Applications in Security
- Artificial Intelligence in Bioinformatics
- Robot Structure Design and Control
- Artificial Intelligence in Biomedical and Healthcare
- Multi-Robot Intelligent Aggregation Mechanisms and Operation Platforms
- Artificial Intelligence and Learning Environments
- Advances in AI-Driven Smart System Designs
- Robot Navigation, Positioning, and Autonomous Control
- Robot Perception and Data Fusion
- Advances in Artificial Intelligence research
- Advances in AI-Driven Data Analytics and Innovation

- Application of Intelligent Systems for Solving Real-World Problems
- Hybrid Systems Design and Applications Using AI
- Robot Grabbing and Operation
- Robot Behavior Decision and Control
- Robot Motion and Path Planning
- Applications of Intelligent Systems and Computer Vision

For additional information, contact:

Book Series Editor: Subhendu Kumar Pani
Professor & Research Co-ordinator
Dept. of Computer Science and Engineering
Orissa Engineering College, Bhubaneswar, India
Email: pani.subhendu@gmail.com

AAP ADVANCES IN ARTIFICIAL INTELLIGENCE & ROBOTICS

Advancements in Artificial Intelligence, Blockchain Technology, and IoT in Higher Education: Mitigating the Impact of COVID-19
Editors: Subhendu Kumar Pani, PhD, Kamalakanta Muduli, PhD, Sujoy Kumar Jana, PhD, Srikanth Bathula, PhD, and Golam Sarwar Khan, PhD

The Fusion of Artificial Intelligence and Soft Computing Techniques for Cybersecurity
Editors: M. Ajabbar, PhD, Sanju Tiwari, PhD, Subhendu Kumar Pani, PhD, and Stephen Huang, PhD

Incorporating AI Technology in the Service Sector: Innovations in Creating Knowledge, Improving Efficiency, and Elevating Quality of Life
Editors: Maria Jose Sousa, PhD, Subhendu Pani, PhD, Francesca dal Mas, PhD, and Sergio Sousa, PhD

Handbook of Research on Artificial Intelligence and Soft Computing Techniques in Personalized Healthcare Services
Editors: Uma N. Dulhare, PhD, A. V. Senthil Kumar, Amit Dutta, PhD, Seddik Bri, PhD, and Ibrahiem M. M. EI Emary, PhD

Fusion of Artificial Intelligence and Machine Learning in Advanced Image Processing
Editors: Arun Rana, PhD, Rashmi Gupta, PhD, Sharad Sharma, PhD, Ahmad A. Elngar, PhD, and Sachin Dhawan, PhD

Artificial Intelligence and Machine Learning for Business
Editors: Namita Mishra, PhD, and A. V. Senthil Kumar, PhD, DSc

Advances in Autonomous Navigation through Intelligent Technologies
Editors: Niharika Singh, PhD, Thipendra P. Singh, PhD, and Brian Azzopardi, PhD

ABOUT THE EDITORS

Arun Kumar Rana, PhD
Department of Electronics and Communication Engineering,
Galgotias College of Engineering & Technology, Greater Noida, India

Arun Kumar Rana, PhD, is currently working as an Assistant Professor at Department of Electronics and Communication Engineering, Galgotias College of Engineering & Technology, Greater Noida, India. He has more than 15 years of experience. His area of interest includes image processing, wireless sensor networks, Internet of Things, AI, machine learning, and embedded systems. He has published more than 100 SCI/ESCI/Scopus-index papers in national and international journals and conferences. He has also published many books with national and international publishers, such as Taylor and Francis, USA. He has attended workshops and faculty development programs and guided six MTech candidates. He serves as a reviewer and editor for several international journals and conferences. He is also a member of the Asia Society of Research. He holds six national and international patents and has conducted many workshops on IoT and its applications in engineering, wireless networks, simulators, etc. He has keen interest in teaching and implementing the latest techniques related to wireless and IoT technology. He has received many international awards from various international organizations.

Rashmi Gupta, PhD
Professor, Electronics and Communication Engineering Department,
Ambedkar Institute of Advanced Communication Technologies and
Research, under Govt. of NCT of Delhi, India

Rashmi Gupta, PhD, currently serves as a Professor in the Electronics and Communication Engineering Department at Ambedkar Institute of Advanced Communication Technologies and Research, under the Government of NCT of Delhi. She also holds the position of Founder Director of the Incubation Center AIACTR-IRF, funded by the Delhi Government. Prior to this, she gained extensive experience in the industry and various academic institutes. Prof. Gupta has authored over 90 research papers in various renowned international journals and conferences. She has published patents and has also been involved in DST-funded projects. Prof. Gupta is associated with several international research organizations as an editorial board member and reviewer. Her primary research interests include machine learning, computer vision, signal, and image processing. She earned her ME and PhD degrees in Electronics and Communication Engineering from Delhi College of Engineering, Delhi University, India.

Sharad Sharma, PhD
Professor, Electronics and Communication Engineering Department, Maharishi Markandeshwar (Deemed to be University), Mullana, India

Sharad Sharma, PhD, currently serves as a Professor in the Department of Electronics and Communication Engineering at Maharishi Markandeshwar (Deemed to be University), Mullana, India. He possesses extensive experience of over 20 years in teaching and administrative roles. With a teaching experience of more than 20 years, he has conducted numerous workshops on topics such as soft computing and its applications in engineering, wireless networks, simulators, etc. He has a strong passion for teaching and implementing the latest techniques in wireless and mobile communications. Additionally, he has established a student chapter of IEEE as a Branch Counselor. His research interests encompass routing protocol design, performance evaluation, and optimization for wireless mesh networks using nature-inspired computing, the Internet of Things, and space communication, among others. He is an active member of various professional organizations. He earned his BTech degree in Electronics Engineering from Nagpur University, Nagpur, India, MTech in Electronics and Communication Engineering from Thapar Institute of Engineering and Technology, Patiala, India, and PhD from National Institute of Technology, Kurukshetra, India.

Ahmed A. Elngar, PhD
Associate Professor and Head of Computer Science Department, Faculty of Computers, and Artificial Intelligence, Beni-Suef University, Egypt

Ahmed A. Elngar, PhD, is an Associate Professor and Head of Computer Science Department, Faculty of Computers, and Artificial Intelligence, Beni-Suef University, Egypt where he is also Director of the Technological and Informatics Studies Center (TISC). Dr. Elngar is the Founder and Head of the Scientific Innovation Research Group (SIRG). Dr. Elngar has more than 55 scientific research papers published in prestigious international journals and over 25 books covering such diverse topics as data mining, intelligent systems, social networks, and smart environment. He is a member of the Egyptian Mathematical Society and International Rough Set Society. His other research areas include Internet of Things (IoT), network security, intrusion detection, machine learning, data mining, artificial intelligence, big data, authentication, cryptology, healthcare systems, and automation systems. He is an editor and reviewer of many international journals around the world. Dr. Elngar has won several awards, including the Young Researcher in Computer Science Engineering Award from Global Outreach Education Summit in 2019.

Sachin Dhawan, PhD
Assistant Professor, Panipat Institute of Engineering and Technology, Samalkha, India

Sachin Dhawan, PhD, is currently working with the Panipat Institute of Engineering and Technology, Samalkha, India, as an Assistant Professor. Prof. Dhawan done his PhD degree in electronics and communication engineering at AIACT&R under Guru Gobind Singh Indraprastha University, Dwarka, New Delhi, India. He was an Assistant Professor at the Geeta Institute of Management and Technology, Kanipla, Kurukshetra, India. He has authored over 20 research papers in various international journals and conferences. He has also published his papers in IEEE conferences and Scopus/SCI-indexed journals. His current research interests include signal and image processing, Internet of Things, artificial intelligence, and machine learning. He is also an associate member of the

All India Institutions of Engineers and a member of the Asia Society of Research. He serves as a reviewer for several journals and international conferences. He has also published two patents. He completed his BTech degree at Kurukshetra University, Kurukshetra, and MTech degree in electronics and communication engineering at the University Institute of Engineering and Technology under Kurukshetra University, Kurukshetra, India.

CONTENTS

CONTRIBUTORS

Alankrita Aggarwal
Panipat Institute of Engineering and Technology, Panipat, Samalkha, Haryana, India

K. N. Agrawal
Project Coordinator, ICAR–Central Institute of Agricultural Engineering, Nabi Bagh, Berasia, Bhopal, India

Anupam Bonkra
Chandigarh Engineering College, Punjab, India

Nidhi Chahal
Chandigarh Engineering College, Punjab, India

N. S. Chandel
Scientist, ICAR–Central Institute of Agricultural Engineering, Nabi Bagh, Berasia, Bhopal, India

Pummy Dhiman
Chitkara University, Punjab, India

M. Shanmuga Eswari
Department of Computer Applications, Kalasalingam Academy of Research and Education, Krishnankoil, Tamil Nadu, India

Shivani Gaba
Panipat Institute of Engineering and Technology, Panipat, Samalkha, Haryana, India

Siddharth Gautam
Department of Computer Science and Engineering, GGSIPU, Delhi, India

Sunita Gond
Faculty, Department of Master of Computer Application, BUIT, Barkatullah University, Bhopal, Madhya Pradesh, India

Ravi Gorli
Assistant Professor, Department of Information Technology, Anil Neerukonda Institute of Technology and Sciences, Visakhapatnam, Andhra Pradesh, India

N. Indumathi
Assistant Professor, Department of Computer Science and Engineering, Rajalakshmi Institute of Technology, Chennai

Rishabh Jha
HOD, Department of Computer Applications, Lumbini Academic College, Tribhuvan University, Kathmandu, Nepal

Priyanshu Joshi
Department of Computer Science and Engineering, GGSIPU, Delhi, India

Sandeep Kajal
College of Engineering Global Campus, Kyung Hee University, Seoul, South Korea

Saurabh Katiyar
Galgotias College of Engineering and Technology, Greater Noida, Uttar Pradesh, India

Amandeep Kaur
Chitkara University, Punjab, India

Inderpreet Kaur
Department of CSE, GCET, Greater Noida, Uttar Pradesh, India

M. V. Kishore
Assistant Professor, Department of Information Technology, Anil Neerukonda Institute of Technology and Sciences, Visakhapatnam, Andhra Pradesh, India

Ram Krishna
Department of Electronics and Communication Engineering, G.L. Bajaj Institute of Technology and Management, Greater Noida, Uttar Pradesh, India

Arun Kumar
Galgotias College of Engineering & Technology, Greater Noida, India

Deepak Kumar
Panipat Institute of Engineering and Technology, Panipat, Samalkha, Haryana, India

Shally Nagpal
Panipat Institute of Engineering and Technology, Panipat, Samalkha, Haryana, India

R. Ramalakshmi
Department of Computer Science and Engineering, Kalasalingam Academy of Research and Education, Krishnankoil, Tamil Nadu, India

Sanjeev Kumar Rana
Department of Computer Science and Engineering, Maharishi Markandeshwar (Deemed to be University), Mullana, Ambala, Haryana, India

Sumit Kumar Rana
Assistant Professor, Panipat Institute of Engineering and Technology, Samalkha, India

Poonam Rani
Department of Computer Science and Engineering, GGSIPU, Delhi, India

Renu
Department of CSE, GCET, Greater Noida, Uttar Pradesh, India

R. Revathy
Department of Computer Applications, Kalasalingam Academy of Research and Education, Krishnankoil, Tamil Nadu, India

Meena Sachdeva
Department of CSE, GCET, Greater Noida, Uttar Pradesh, India

Sudhir Kumar Sharma
Jaipur National University, Jaipur, Rajasthan, India

Anjali Kumari Singh
BSc Nursing, Norvic International Hospital, Kathmandu, Nepal

Poonam Sinha
Head, Department of Electronics and Communication Engineering, BUIT, Barkatullah University, Bhopal, Madhya Pradesh, India

Priya Trivedi
Aisect College of Professional Studies, Indore, Madhya Pradesh, India

Nitin Tyagi
Department of Computer Science and Engineering, GGSIPU, Delhi, India

Aditi Upadhyay
Jaipur National University, Jaipur, Rajasthan, India

Shrikant Upadhyay
Cambridge Institute of Technology, Bangalore, Karnataka, India

K. Upendar
PhD Research Scholar, ICAR–Central Institute of Agricultural Engineering, Nabi Bagh, Berasia, Bhopal, India

PREFACE

This book focuses on the fusion of artificial intelligence and machine learning in advanced image processing, data analysis, and cybersecurity. It compiles and discusses various engineering solutions using different artificial intelligence paradigms. The book explores recent technological advancements in these fields and considers how artificial intelligence, machine learning, deep learning, soft computing, and evolutionary computing techniques can be used to design, implement, and optimize engineering solutions. It aims to provide readers with the insight required to use digital imaging tools to solve new problems.

The book is organized as follows:

Chapter 1 provides an overview of AI- and IoT-based medical imaging technology for healthcare applications, including its needs, benefits, challenges, existing structure, and possible future technologies.

Chapter 2 discusses the early diagnosis of cyanotic congenital heart disease using deep learning and details the associated benefits and challenges.

Chapter 3 covers the issues and challenges of AI- and IoT-based smart healthcare with image processing. It also discusses advances in fault ride-through capability with solutions.

Chapter 4 discusses cyber-physical systems in medical science, describing control strategies for advanced healthcare systems in detail.

Chapter 5 focuses on the significance of machine learning (ML) and natural language processing (NLP) in sentiment analysis, which is used to improve data prediction capability.

Chapter 6 presents intelligent innovative systems, offering a multi-disciplinary outlook on current challenges and contingencies for grid applications.

Chapter 7 describes the usage of cloud computing and IoT in artificial intelligence.

Chapter 8 deals with the basics of a blockchain-based model to preserve the authenticity of judicial evidence. It includes a detailed case study on preserving the authenticity of the judicial evidence system.

Chapter 9 describes various applications of machine learning and deep learning techniques in agriculture for weeds and crop disease detection and classification.

Chapter 10 covers the basics and structure of video image identification and verification using a preliminary score approach for authentication and security enhancement.

Chapter 11 focuses on machine learning and cyber defense, examining the hype versus reality.

Chapter 12 describes an effective investigation of security and privacy in vehicular ad-hoc networks with preventive measures.

Finally, Chapter 13 discusses various challenges and research perspectives in the automation of image processing with artificial intelligence and pattern recognition methods.

The editors and contributing authors hope that readers will benefit from and gain a basic understanding of artificial intelligence and machine learning for advanced image processing, data analysis, and cybersecurity. This book aims to enhance readers' knowledge and make an impact in realizing future emerging technologies worldwide in the near future.

PART I

Artificial Intelligence and Machine Learning in Medicine and Healthcare

CHAPTER 1

AI AND IOT-BASED MEDICAL IMAGING TECHNOLOGY FOR HEALTHCARE APPLICATIONS: FUTURE RESEARCH

SANDEEP KAJAL,[1] SAURABH KATIYAR,[2] RAM KRISHNA,[3] and ARUN KUMAR[4]

[1]*College of Engineering Global Campus, Kyung Hee University, Seoul, South Korea*

[2]*Galgotias College of Engineering and Technology, Greater Noida, Uttar Pradesh, India*

[3]*Department of Electronics and Communication Engineering, G.L. Bajaj Institute of Technology and Management, Greater Noida, Uttar Pradesh, India*

[4]*Galgotias College of Engineering and Technology, Greater Noida, Uttar Pradesh, India*

ABSTRACT

Recently, there has been a lot of interest in using artificial intelligence (AI) and the internet of things (IoT) in radiography. Although part of this interest may have been sparked by excessive hopes that the technology might perform some jobs better than radiologists, there is mounting proof of its limits in medical imaging. IoT and AI are a perfect option to provide the uniformity, consistency, and reliability needed to enable radiologists in

The Fusion of Artificial Intelligence and Soft Computing Techniques for Cybersecurity.
M. A. Jabbar, Sanju Tiwari, Subhendu Kumar Pani, & Stephen Huang (Eds.)

their quest to provide superior patient care due to the unbounded potential of computers. The IoT in medical imaging would cut down on waiting times and frustration for both patients and doctors. As digitization spread to many sectors of medical technology, it also cleared the way for the monitoring and administration of medical equipment. In order to demonstrate the importance of the IoT and AI in medical imaging technology, the chapter reviews IoT and AI-based medical imaging technology for use in healthcare applications.

1.1 INTRODUCTION

It is not new to combine the internet of things (IoT) and artificial intelligence (AI), but recent advancements in computing power, data accessibility, and algorithm performance have sparked an "AI renaissance" in which AI is being applied to almost every industry with the goal of maximizing productivity, efficiency, and accuracy. An unprecedented degree of interest in IoT and AI was prompted by reports that deep learning (DL) algorithms may perform at human-levels in a handful of specialized vocations [1]. Artificial neural networks (ANNs) can be used for various non-medical purposes, such as identifying cursive handwriting, driverless cars, optical telescope focusing, and many more. Figure 1.1 showcases a few IoT and AI applications in the medical field.

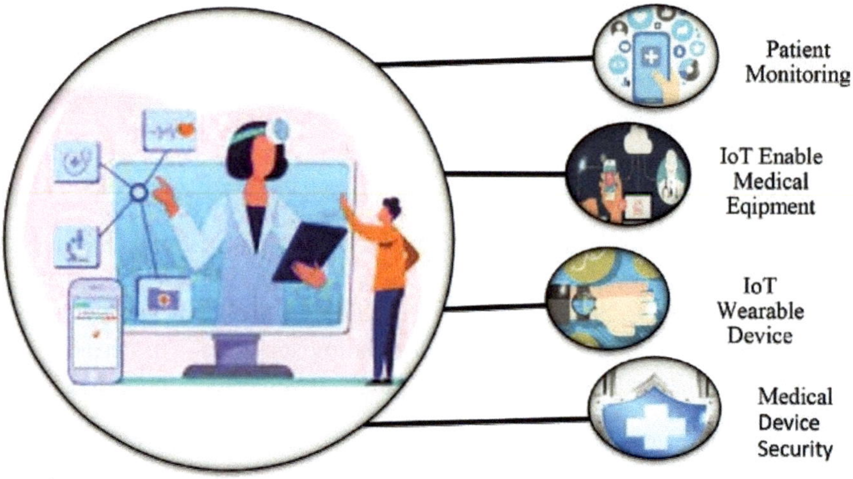

FIGURE 1.1 Applications of IoT and AI in Healthcare.

The expansion of the medical sector and the creation of techniques like computed tomography (CT) and magnetic resonance imaging (MRI) produce massive volumes of high-dimensional and variable-rich data every day [2–5]. As a result, the dimensions of medical imaging databases are rapidly growing. Managing the file system with the growing data volume resulting from the development in medical databases is challenging. The management of medical data has been a top issue for healthcare service providers in the United States [6–9]. Consequently, cloud computing has been utilized in the medical industry to store and process medical data due to its simplicity in usage for medical images. Cloud computing provides scalable and adaptable computer resources from remote locations according to the user's needs [10]. It is also effective in delivering computational resources in the high-end computer context [11]. Since the cloud platform is used for data storage in cloud computing, customers do not require their own storage or servers on their devices.

As a result, consumers will not need to have a lot of servers or storage space in their PCs [12]. Therefore, cloud computing and long-term medical picture record storage are effective for resolving a number of issues in the medical industry [13]. Cloud computing is widely employed in the medical industry to store, process, and share patient medical records. In a cloud computing environment, the hospital simply needs to gather patient data from the files and send it to the cloud for storage. In this situation, AI is useful as it can swiftly infer conclusions from the data. Machine learning (ML) systems may provide operational forecasts up to 20 times earlier and more accurately than traditional business intelligence tools, which typically watch for numeric thresholds to be passed [14].

The term IoT refers to a world that is completely interconnected [15]. It connects everything through technology, creating a new universe where kids can interact with each other through the internet. IoT is more than just an idea; it has the potential to revolutionize technology and transform people's lifestyles [16]. The IoT will impact any physical or theoretical idea, as it requires object communication. Anyone should be able to access any material from any device at any time, from anyone located anywhere, who is part of any company or service, over any path or network. In a positive sense, 'availability' is a crucial component that influences IoT performance [17–19].

1.1.1 MOTIVATION, OBJECTIVES, AND CONTRIBUTIONS

Sensors and embedded devices in vehicles, phones, watches, supermarkets, houses, highways, and bridges, appliances, industrial and farm equipment, and wearable technology are already enabling new types of information and altering how it is created, consumed, and experienced. IoT and AI provide a huge potential for data analytic advancements. The links between IoT, data storage and processing, and ML are clear and are already gaining traction. The following are the major objectives:

- This chapter presents a review of IoT and AI in terms of energy and security requirements for medical imaging applications.
- To discuss sources of the medical imaging available in the medical industry.
- To explain the IoT and AI vision and terminology.
- To facilitate a debate on the concerns and difficulties of open IoT and AI that will be addressed by future research.
- To elaborate on the five-layer IoT architectures.

1.2 LITERATURE REVIEW

Medical data is rapidly expanding due to the current innovation in healthcare services. Healthcare services in the medical field encompass the diagnosis, treatment, therapy, and prevention of illnesses, injuries, and mental health conditions. The quality of these healthcare sectors relies on efficient problem-solving and resource allocation, which necessitate the application, collection, and management of medical data [21]. Prior to the advent of modern technologies, many organizations relied on handwritten patient records organized in the form of reports [22]. However, modern technology is now essential for analyzing and managing large datasets. IoT and cloud computing have been adopted to efficiently handle, store, process, and transmit medical data. Cloud computing offers a flexible and dynamic approach to managing medical data, while the IoT enables connectivity between devices, applications, data, objects, and sensors, significantly impacting the data transmission process in the medical industry [23]. In a cloud computing setting, IoT enables continuous interaction and communication between devices and sensors.

Because it makes operations more convenient, more organizations are now starting to use IoT and cloud computing technologies [24]. Additionally, organizations engaged in pharmaceutical and medical research are using cloud computing and IoT, and their applications have the potential to significantly improve society. Convolutional neural networks (CNNs) have the most potent architecture among DL techniques, and they offer superior classification accuracy [25]. CNN is a DL technique used to address challenging classification issues [26]. For large datasets, it enhances computing efficiency and accuracy rate. Additionally, it automatically extracts feature maps linked to training data.

The model is pretrained with relevant and adequate data throughout the classification process, after which the model's feature description is assigned. As a result, instead of starting with random variables, the model begins with the pattern that is associated with the task. The CNN cannot perform well without relevant data, so the learning process still requires training datasets that are connected to testing datasets [27]. To simulate nonlinear functions, CNNs are combined with fuzzy logic systems to create fuzzy CNNs. Fuzzy CNNs improve the effect of function approximation.

1.3 FIVE LAYER ARCHITECTURE

The 5-layer architecture is considered the most effective when working on projects that involve a range of cutting-edge technologies and diverse application areas [28]. As shown in Figure 1.2, the five-layer model includes two additional levels compared to the basic model, making it an extension of the fundamental design of IoT.

1. **Perception Layer:** The first layer of IoT architecture is the perception layer. It consists of a multitude of sensors and actuators that are used to collect valuable information such as temperature, moisture content, intrusion detection system (IDS), noises, and more. The main purpose of this layer is to gather data from the environment and transmit it to another layer, where actions can be taken based on that data.

2. **Network Layer:** It is the layer that connects the perceptual and middleware layers, as the name implies. It uses networking technologies like 3G, 4G, UMTS, Wi-Fi, infrared, and others to obtain

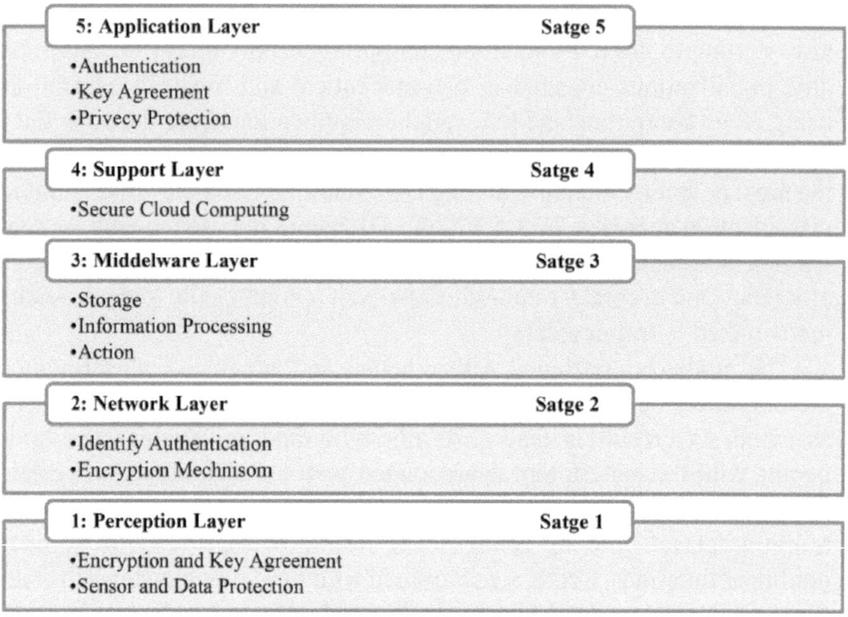

FIGURE 1.2 Five-layer architecture of internet of things.

data from the perception layer and send it to the middleware layer. Because it is responsible for communication between the perception and middleware layers, it is also known as the communication layer. All data transfers are done securely, with the collected information remaining confidential.

3. **Middleware Layer:** This includes sophisticated functionality such as storage, calculation, processing, and action-taking. It saves all data and sends it to the relevant device depending on the device's address and name. It may also make judgments based on computations performed on sensor data sets.

4. **Business/Support Layer:** Any device's success is determined not only by the technology employed in it but also by how it is presented to its users. These responsibilities are managed by the device's business layer, which includes creating flowcharts, graphs, analyzing data, and determining how the device can be enhanced, among other tasks.

5. **Application Layer:** All application activities are handled by the application layer using data obtained from the middleware layer. This application can be used to perform a variety of functions, including sending emails, setting alarms, turning on or off security systems, wearables, and smart agriculture.

1.4 MEDICAL IMAGING SOURCES

In order to diagnose, monitor, or treat medical disorders, the term "medical imaging" refers to a number of various technologies that are used to visualize the human body. Each technology type provides distinct information on the part of the body being investigated or treated, in relation to potential illness, damage, or the efficacy of medical care. The connectivity of the IoT allows medical imaging equipment to operate more efficiently and effectively, while reducing costs. This section provides details on the various sources of medical imaging and how the IoT contributes to illness detection through medical imaging, as depicted in Figure 1.3. The limited sources of medical imaging used by the medical community to diagnose diseases are also detailed in this section [29].

1. **Ultrasound Imaging (Sonography):** High-frequency sound waves are used in ultrasound imaging, also known as sonography, to look within the body. Real-time ultrasound imaging allows for the display of blood flowing via blood vessels as well as movement of the body's interior organs. Contrary to X-ray imaging, ultrasonic imaging does not expose patients to ionizing radiation.

   A transducer (probe) is positioned either on the skin or within a bodily orifice during an ultrasonic examination. The skin is covered with a small layer of gel to allow the ultrasonic waves to pass through the gel and into the body.

   Figure 1.4 depicts the probe (transducer) used during an ultrasound examination. The ultrasound picture is created as a result of the waves' reflection off the body's structures. The information required to create a picture is provided by the intensity (amplitude) of the sound signal and the duration of the wave's passage through the body.

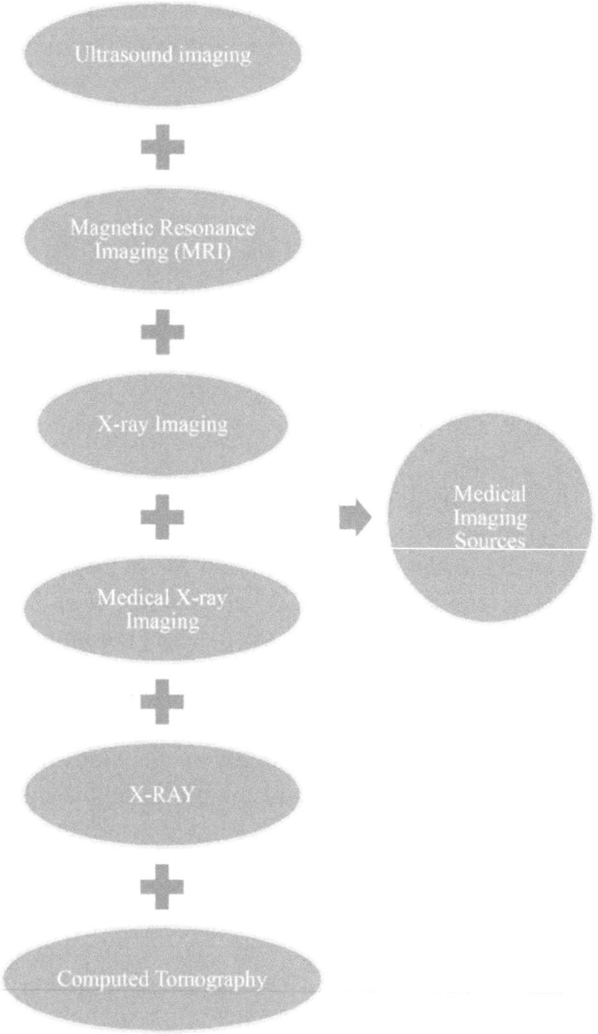

FIGURE 1.3 Medical imaging sources.

2. **Magnetic Resonance Imaging (MRI):** A medical imaging process called MRI creates pictures of the inside body structures. To create pictures, MRI scanners employ radio waves (radiofrequency energy) and high magnetic fields. Protons found in water and fat molecules make up the majority of the signal in MR images.

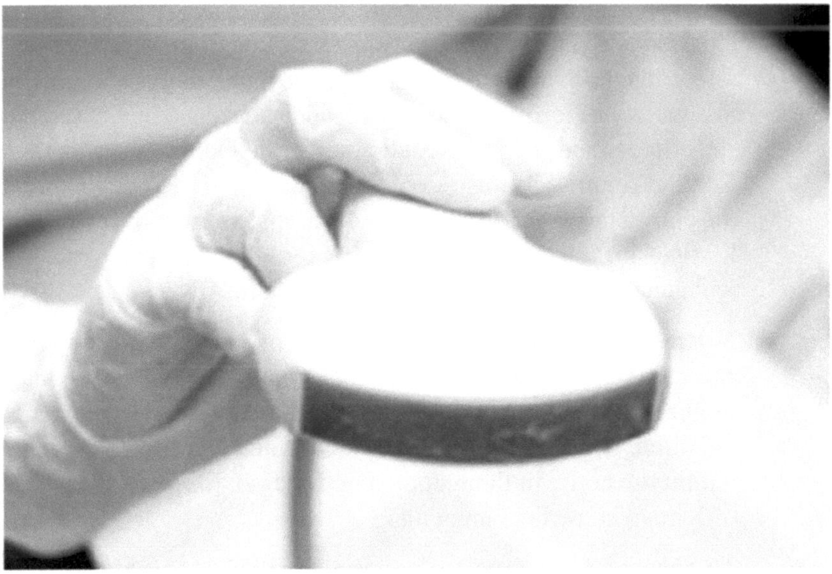

FIGURE 1.4 Transducer (probe) used during an ultrasound exam [20].

Source: Degum. https://www.degum.de/presse/pressemitteilungen/im-detail/news/bauchultraschall.html

An electric current is sent via coils of wire during an MRI test to produce a transient magnetic field inside the body of the patient. A transmitter and receiver in the device broadcast and receive radio waves, and these signals are utilized to create digital photographs of the body part being scanned. Depending on the portion of the body being examined, a normal MRI scan might take anywhere between 20 and 90 minutes [30].

3. **X-Ray Imaging for Pediatrics:** The identification and management of a variety of pediatric medical disorders have improved due to medical X-ray imaging. Pediatric patients are those who are under the age of 21 at the time of their diagnosis or treatment, according to the Federal Food, Drug, and Cosmetic Act (FD&C Act). They are typically divided into several categories based on age ranges (neonates, infants, children, and adolescents). The size of the juvenile patient is even more crucial to consider for medical X-ray imaging than the patient's age, as the patient's size affects the amount of radiation required to produce a high-quality medical image [31].

4. **Medical X-Ray Imaging:** Numerous medical disorders in both children and adults can now be diagnosed and treated more accurately, thanks to medical imaging. Medical imaging processes come in a variety of forms, or modalities, and each one employs a unique set of tools and methods. Radiography, including mammography, CT, and fluoroscopy, uses ionizing radiation to provide pictures of the body. However, the chance of developing cancer over a person's lifetime may increase if they are exposed to ionizing radiation, which has enough energy to potentially harm DNA.

5. **X-Ray (X-Radiation):** The utilization of radiation allows for the sharp, precise imaging provided by X-rays. X-rays create images of the internal structure by penetrating it with electromagnetic radiation, which is useful for diagnosing and identifying anomalies within the body. In their study, Kieffer et al. [32] describe the use of X-rays in medical imaging.

   To enhance the contrast of certain MRI scans, intravenous (IV) medications such as gadolinium-based contrast agents (GBCAs) are used. GBCAs, which are rare earth metal-based, are typically administered through an IV in the arm.

6. **Computed Tomography (CT):** The non-invasive medical treatment known as CT, sometimes referred to as "computerized tomography" or "computed axial tomography" (CAT), creates cross-sectional images of the body using specialized X-ray equipment. Similar to the slices in a loaf of bread, each cross-sectional image displays a "slice" of the individual being examined. These cross-sectional images have various diagnostic and therapeutic applications.

   CT scans can be performed on any part of the body for multiple purposes, such as diagnosis, treatment planning, intervention, or screening. The majority of CT scans are conducted as outpatient procedures.

1.5 IOT WITH ARTIFICIAL INTELLIGENCE (AI)

Applications and installations for the IoT are increasingly utilizing AI. Investments and acquisitions in companies that combine AI and IoT have grown over the past two years. Major IoT platform software companies currently offer integrated AI features, including ML-based analytics. Information that previously required human analysis can now be extracted from data using other AI technologies such as speech recognition and computer vision. IoT AI applications help organizations reduce unscheduled downtime, enhance operational effectiveness, create new products and services, and better manage risk. AI-enabled IoT can also improve operational efficiency. By processing continuous streams of data to identify patterns that are not visible to the human eye or evident on basic gauges, ML can predict operational conditions and identify parameters that can be adjusted in real-time to maintain optimal results [33].

1.6 FUTURE RESEARCH

IoT and AI applications in the medical field have a long history of development. As a result, several ongoing issues and difficulties in the medical sector have also compelled numerous research teams to attentively examine AI. AI technology is now being used more widely in the medical industry due to the introduction of cutting-edge technologies such as the IoT, cloud computing, big data, and 5G mobile networks. Furthermore, the complete integration of AI and IoT technology allows for the continuous advancement of medical diagnosis and treatment capacities, effectively serving the general populace. Future medical equipment, including those used in operating rooms, critical care units, emergency rooms, wards, and even wearable community and family health devices, will progressively be connected to IoT terminals to generate vast amounts of medical data. Edge computing, with its outstanding practicability, deploy ability, and usability, is incorporated into the IoT access terminal of medical equipment for localized data processing, storage, and edge intelligent services. IoT terminal devices for medical equipment enabled with edge computing will steadily advance in the near future.

The main objective of this chapter is to provide a comprehensive overview of unresolved issues and challenges in the fields of medical imaging in IoT and AI. In this section, we present some background information on emerging issues and ideas that will be further researched. According to recent contributions and research publications, the majority of challenges arise from increasing traffic demands with various types of traffic, greater variation in data formats, diverse IoT and AI devices, heterogeneous networks, and so on. Each of these elements significantly impacts the effectiveness of IoT and AI systems. Specific application needs exist in terms of computing and communication capabilities, quality of service (QoS), mobility, reliability, privacy and security, and other factors. IoT and AI applications rely on other systems, such as cloud computing, to meet these needs. The architecture of IoT and AI systems may include additional cloud-based technologies like fog computing, cloudlets, MEC, and MCC [34]. The majority of IoT and AI apps and services available today are cloud-based, utilizing data exchange between smart devices and cloud infrastructure to provide services. This approach faces several challenges, including technological challenges (such as network latency, throughput, and reliability), financial challenges (such as the cost of data transfer to the Cloud), and social and administrative challenges.

Fog computing, cloudlets, MEC, MCC, and other architectures and computing paradigms have all been introduced in the past year to deploy infrastructure closer to the source of data. Different models are distinguished by their "infrastructure distance" from IoT and AI components, the kind of infrastructure where functions are allocated, and other elements. The storage and computational power of network edge equipment is inferior to that of cloud computing. Long-term projects like data storage and pattern analysis are best suited for it. It might be based on an algorithm that allows the load allocation among IoT and AI subsystems to be optimized. It may be feasible to offer a categorization strategy for IoT and AI applications based on their performance profile (e.g., QoS needs) utilizing this type of assessment. This kind of assessment might also be used to compare alternative application architectures and resource management policies. This approach might also be used to evaluate QoS and QoE, as well as to simulate IoT and AI systems. IoT and AI system mathematical formulation and assessment approaches that are adequate can break through future progress [35].

In the future, IoT and AI will be involved in every facet of hospital administration and care. The accuracy of inventory and materials is increased thanks to IoT, electronic tags specifically designed for medical equipment assets based on QR codes, RFID, and other technologies. Real-time monitoring of equipment status is made possible through the use of IoT sensing terminals. Importantly, this will support the achievement of thorough and unbiased management of medical equipment. The collection, categorization, packaging, and storage of medical waste in hospitals and other healthcare facilities, as well as the subsequent treatment and management of medical waste, have all benefited from the oversight and control of the entire process.

1.7 CONCLUSION

For more than three decades, clinical research, translation, and practice have been fueled by medical imaging informatics. The advancements in related fields of research that this chapter highlights hold the possibility of revolutionizing imaging informatics as it is now used throughout the healthcare continuum by enabling well-informed, more accurate diagnosis, prompt prognosis, and efficient treatment planning. Medical imaging informatics makes up a sizable portion of AI-based research-driven techniques that have been approved by the Food and Drug Administration (FDA). To deliver intelligent services, IoT and AI must enable seamless communication anytime, everywhere, by anyone and anything. This includes identifying, sensing, networking, processing, and visualization capabilities. This notion opened up a slew of new options for large-scale service and product development, resulting in a flood of new ideas and commercial prospects. New IoT and AI applications, along with developing technologies, present new challenges that the research community needs to address. We have summarized and analyzed these challenges, as well as described the major obstacles, by envisioning future research topics and advancements that are most intriguing. A brief overview of interesting research subjects, a classification of significant concerns based on specific functional domains, and insights into specific emerging challenges and ideas all contribute to facilitating future research. Consequently, this work can serve as a valuable starting point for future studies, benefiting the research community. Handling, indexing, querying, and analyzing digital pathology data pose challenges for imaging researchers. Due to practical storage and processing limitations, it is not feasible to comprehensively compare query data with each sample in a high-dimensional database. Managing relatively large-scale, multi-dimensional datasets that will continue to grow over time is one of the key issues. The second challenge lies in accurately examining the features of data originating from various modalities.

KEYWORDS

- **artificial intelligence**
- **automation**
- **healthcare**
- **internet of things**
- **medical imaging**
- **resource optimization**

REFERENCES

1. Alseelawi, N., Hazim, H. T., & Salim, A. H. T., (2022). A novel method of multimodal medical image fusion based on hybrid approach of NSCT and DTCWT. *International Journal of Online & Biomedical Engineering, 18*(3).
2. Ibrahim, R. W., Jalab, H. A., Karim, F. K., Alabdulkreem, E., & Ayub, M. N., (2022). A medical image enhancement based on generalized class of fractional partial differential equations. *Quantitative Imaging in Medicine and Surgery, 12*(1), 172.
3. Yu, Q., et al., (2022). Crosslink-net: Double-branch encoder network via fusing vertical and horizontal convolutions for medical image segmentation. *IEEE Transactions on Image Processing.*
4. Behera, T. M., Mohapatra, S. K., Samal, U. C., Khan, M. S., Daneshmand, M., & Gandomi, A. H., (2019). Residual energy-based cluster-head selection in WSNs for IoT and AI application. *IEEE Internet of Things Journal, 6*(3), 5132–5139.
5. Behera, M., Samal, U., & Mohapatra, K., (2018). Energy efficient modified LEACH protocol for IoT and AI applications. *IET Wireless Sensors System, 8*(5), 223–228.
6. Tsiknas, K., Taketzis, D., Demertzis, K., & Skianis, C., (2021). Cyber threats to industrial IoT and AI: A survey on attacks and countermeasures. *IoT and AI, 2*(1), 163–188.
7. Ojanperä, T., Mäkelä, J., Majanen, M., Mämmelä, O., Martikainen, O., & Väisänen, J., (2021). Evaluation of LiDAR data processing at the mobile network edge for connected vehicles. *EURASIP Journal on Wireless Communications and Networking,* (1), 1–23.
8. Rana, A. K., & Sharma, S., (2021). Industry 4.0 manufacturing based on IoT and AI, cloud computing, and big data: Manufacturing purpose scenario. In: *Advances in Communication and Computational Technology* (pp. 1109–1119). Springer, Singapore.
9. Oktian, Y. E., Witanto, E. N., & Lee, S. G., (2021). A conceptual architecture in decentralizing computing, storage, and networking aspect of IoT and AI infrastructure. *IoT and AI, 2*(2), 205–221.

10. Almezhghwi, K., Serte, S., & Al-Turjman, F., (2021). Convolutional neural networks for the classification of chest X-rays in the IoT and AI era. *Multimedia Tools and Applications*, 1–15.
11. Bhushan, B., Khamparia, A., Sagayam, K. M., Sharma, S. K., Ahad, M. A., & Debnath, N. C., (2020). Blockchain for smart cities: A review of architectures, integration trends and future research directions. *Sustainable Cities and Society, 61*, 102360.
12. Bhushan, B., Sahoo, C., Sinha, P., & Khamparia A., (2021). Unification of blockchain and internet of things (BIoT and AI): Requirements, working model, challenges, and future directions. *Wireless Networks, 27*(1), 55–90.
13. Inam, H., & Al-Turjman, F., (2021). Intelligent free energy usage through radiant energy space phenomenon: An IoT and AI-powered prototype for modified Bedini generator. *Microprocessors and Microsystems*, 104319.
14. Jamil, H., Umer, T., Ceken, C., & Al-Turjman, F., (2021). Decision-based model for real-time IoT and AI analysis using big data and machine learning. *Wireless Personal Communications*, 1–13.
15. Berntzen, L., Johannessen, M. R., & Florea, A., (2016). Sensors and the smart city: Creating a research design for sensor-based smart city projects. In: *ThinkMind// SMART, The Fifth International Conference on Smart Cities, Systems, Devices and Technologies.*
16. Lin, K., Chen, M., Deng, J., Hassan, M. M., & Fortino, G., (2016). Enhanced fingerprinting and trajectory prediction for IoT and AI localization in smart buildings. *IEEE Transactions on Automation Science and Engineering, 13*(3), 1294–1307.
17. Alletto, S., Cucchiara, R., Del Fiore, G., Mainetti, L., Mighali, V., Patrono, L., & Serra, G., (2015). An indoor location-aware system for an IoT and AI-based smart museum. *IEEE Internet of Things Journal, 3*(2), 244–253.
18. Kaur, M. J., & Maheshwari, P., (2016). Building smart cities applications using IoT and AI and cloud-based architectures. In: *2016 International Conference on Industrial Informatics and Computer Systems (CIICS)* (pp. 1–5).
19. Sun, Y., Wu, T. Y., Zhao, G., & Guizani, M., (2014). Efficient rule engine for smart building systems. *IEEE Transactions on Computers, 64*(6), 1658–1669.
20. McAuliffe, M. J., Francois, M. L., McGarry, D., William, G., Karl, C., & Benes, L. T., (2001). Medical image processing, analysis, and visualization in clinical research. In: *Proceedings 14th IEEE Symposium on Computer-Based Medical Systems. CBMS 2001* (pp. 381–386). IEEE.
21. Scholl, I., Aach, T., Deserno, T. M., & Kuhlen, T., (2011). Challenges of medical image processing. *Computer science-Research and Development, 26*(1), 5–13.
22. Zhang, X., Pipattanasomporn, M., Chen, T., & Rahman, S., (2019). An IoT and AI-based thermal model learning framework for smart buildings. *IEEE Internet of Things Journal, 7*(1), 518–527.
23. Tournier, J. D., Smith, R., Raffelt, D., Tabbara, R., Dhollander, T., Pietsch, M., & Connelly, A., (2019). MRtrix3: A fast, flexible, and open software framework for medical image processing and visualization. *Neuroimage, 202*, 116137.
24. Maier, A., Syben, C., Lasser, T., & Riess, C., (2019). A gentle introduction to deep learning in medical image processing. *Zeitschrift für Medizinische Physik, 29*(2), 86–101.

25. Bhattacharya, S., Maddikunta, P. K. R., Pham, Q. V., Gadekallu, T. R., Chowdhary, C. L., Alazab, M., & Piran, M. J., (2021). Deep learning and medical image processing for coronavirus (COVID-19) pandemic: A survey. *Sustainable Cities and Society, 65,* 102589.
26. Jasti, V., Zamani, A. S., Arumugam, K., Naved, M., Pallathadka, H., Sammy, F., & Kaliyaperumal, K., (2022). Computational technique based on machine learning and image processing for medical image analysis of breast cancer diagnosis. *Security and Communication Networks, 2022.*
27. Dhiman, G., Juneja, S., Viriyasitavat, W., Mohafez, H., Hadizadeh, M., Islam, M. A., & Gulati, K., (2022). A novel machine-learning-based hybrid CNN model for tumor identification in medical image processing. *Sustainability, 14*(3), 1447.
28. Debauche, O., Mahmoudi, S., & Moussaoui, Y., (2020). Internet of things learning: A practical case for smart building automation. *International Conference on Cloud Computing and Artificial Intelligence: Technologies and Applications.* Marrakech, Morocco.
29. Pierleoni, P., Conti, M., Belli, A., Palma, L., Incipini, L., Sabbatini, L., & Concetti, R., (2019). IoT and AI solution based on MQTT protocol for real-time building monitoring. In: *IEEE 23rd International Symposium on Consumer Technologies (ISCT)* (pp. 57–62).
30. Alezabi, K. A., Hashim, F., Hashim, S. J., Ali, B. M., & Jamalipour, A., (2020). Efficient authentication and re-authentication protocols for 4G/5G heterogeneous networks. *EURASIP Journal on Wireless Communications and Networking*, 1–34.
31. Hajjaji, Y., Boulila, W., Farah, I. R., Romdhani, I., & Hussain, A., (2021). Big data and IoT and AI-based applications in smart environments: A systematic review. *Computer Science Review, 39*, 100318.
32. Chegini, H., Naha, R. K., Mahanti, A., & Thulasiraman, P., (2021). Process automation in an IoT and AI–fog–cloud ecosystem: A survey and taxonomy. *IoT and AI, 2*(1), 92–118.
33. Zhang, Y., Sun, Y., Jin, R., Lin, K., & Liu, W., (2021). High-performance isolation computing technology for smart IoT and AI healthcare in cloud environments. *IEEE Internet of Things Journal,* 1–10.
34. Jacob, T. P., Pravin, A., Ramachandran, M., & Al-Turjman, F., (2021). Differential spectrum access for next-generation data traffic in massive-IoT and AI. *Microprocessors and Microsystems, 82*, 103951.
35. Kuwahara, Y., Aihara, N., Yamazaki, S., Ohuchi, K., & Mizuno, H., (2021). Energy-efficiency comparison of Ad-Hoc routings in a shadowing environment for smart IoT and AI. *International Conference on Information Networking (ICOIN).* (pp. 801–804). IEEE.

CHAPTER 2

EARLY DIAGNOSIS OF CYANOTIC CONGENITAL HEART DISEASE USING DEEP LEARNING

RISHABH JHA[1] and ANJALI KUMARI SINGH[2]

[1]*HOD, Department of Computer Applications, Lumbini Academic College, Tribhuvan University, Kathmandu, Nepal*

[2]*BSc Nursing, Norvic International Hospital, Kathmandu, Nepal*

ABSTRACT

Congenital heart disease (CHD) is the special condition of defects in the normal functioning of heart from the time of the birth of an infant. According to the NHS UK [1], CHDs are affecting up to 1 out of the 100 births in UK. Similarly, according to the CDC USA [1], about 1% of the births are affected by the CHDs every year that amounts to nearly 40,000 cases in USA alone per year. Due to the advancements in computer science, mysteries of medical sciences are constantly being uncovered. In all the domains of medical sciences like diagnosis, prognosis, and treatment, machine learning (ML) and deep learning (DL) are highly contributing by adding a clear picture about the condition with collaboration of medical experts. This chapter is proposing a novel technique that uses DL to diagnose cyanotic congenital disease at an early phase so as to minimize the risk of deaths due to the CHDs. This proposed research is based on convolutional neural networks (CNNs), a DL algorithm which is to be employed on the data of SpO_2 of the potential infants by the use of smart wearables.

The Fusion of Artificial Intelligence and Soft Computing Techniques for Cybersecurity.
M. A. Jabbar, Sanju Tiwari, Subhendu Kumar Pani, & Stephen Huang (Eds.)

2.1 INTRODUCTION

Cardiovascular diseases are one of the major causes of death in people from developing and underdeveloped countries. Although most CVDs have advanced infrastructure treatments, late diagnosis often leads to the development of severe symptoms and ultimately, fatalities. Early diagnosis can lead to complete cure in most cases, but the current medical infrastructure lacks the capability to predict the possibility of these diseases, especially congenital heart diseases (CHDs) and rheumatic heart diseases that manifest from birth or early childhood. For at-risk populations, remote health monitoring is a useful tool for providing preventative care and early intervention. Thanks to recent improvements in Internet-of-Things (IoT) concepts, such monitoring systems are now available, enabling omnipresent monitoring [11–16]. Given the critical condition of the patients being monitored, these systems require a high level of quality in features such as availability and accuracy. Deep learning (DL) algorithms show great promise in health applications where a large amount of data is accessible, achieving satisfactory performance [2, 17–26]. This chapter will focus on the use of DL techniques to address the issue of late diagnosis of CHDs.

2.1.1 CONGENITAL HEART DISEASE (CHD)

Congenital heart diseases (CHDs) are a type of cardiovascular disease that occurs before birth in the fetus, resulting in structural damage and malfunctioning of the heart. It may be related to abnormalities of the intrathoracic great vessels. CHDs are the most common cause of death among other congenital diseases. Although the exact causes of CHDs are not yet known, there are multiple factors that influence their development. It may result from a combination of genetic predisposition and environmental factors. In terms of genetic causes, chromosomal abnormalities, mainly aneuploidies, are the primary factors. Trisomy 21, 13, and 18, as well as Turner's syndrome, are contributing factors that result in structural deformations of the heart during fetal development, ultimately leading to CHDs. Environmental factors include maternal illness, phenylketonuria, exposure to drugs, and viral infections during pregnancy.

2.1.2 SIGN AND SYMPTOMS

- Bluish or purple discoloration of skin (cyanosis);
- Rapid heartbeat (tachycardia);
- Rapid breathing (tachypnea);
- Swelling of legs, tummy or around the eyes;
- Passive behavior of children;
- Pulmonary edema.

2.1.3 TYPES

Based on various causes and resulting conditions, CHDs are broadly divided into two types:

i. **Cyanotic CHDs:** These result in bluish skin discoloration due to a lack of oxygen. This category includes a group of lethal CHDs such as pulmonary atresia, tricuspid atresia, truncus arteriosus, and hypo plastic left heart.

ii. **Acyanotic CHDs**: These are those that result in congestive heart failure. They include ventricular septal defects, atrial septal defects, patent ductus arteriosus, etc.

2.1.4 DEVELOPMENT OF CONGENITAL HEART DISEASES (CHDS) IN ADULTS

Without proper diagnosis and care, CHDs may lead to life threatening conditions in adults. Some of these conditions are:

1. **Arrhythmia:** It means irregular heartbeats that occur when the electrical signals coordinating our heartbeat do not work properly. In this condition, the heart beats either too fast, too slow, or irregularly. In severe cases, it may cause a stroke or sudden cardiac death if left untreated.
2. **Endocarditis:** It is the infection of the inner lining of the heart, known as the endocardium. If left untreated, it can lead to heart valve damage or even stroke.

3. **Stroke:** In the development of CHDs in adults, blood clots may travel to the human brain through the bloodstream, resulting in a reduced supply of oxygen to the brain and ultimately leading to a stroke.
4. **Heart Failure:** In certain conditions, CHDs may lead to congestive heart failure, in which the heart cannot pump enough blood to meet the body's needs, ultimately resulting in death.
5. **Unsuccessful Pregnancy:** Some complex CHD cases may lead to an unsuccessful pregnancy. Furthermore, both men and women with CHDs are at risk of passing their disease in some form to their children.

2.1.5 DEEP LEARNING (DL)

Deep learning (DL) is a subset of machine learning (ML) which is in turn a subset of artificial intelligence (AI) as shown in Figure 2.1.

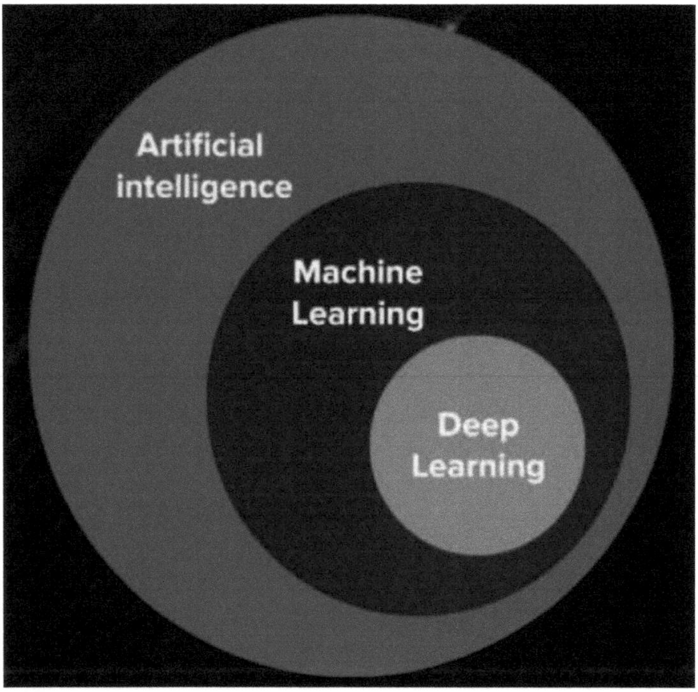

FIGURE 2.1 AI vs. ML vs. DL.

DL draws inspiration from the way human neurons form a learning network, training machines to learn new concepts. The fundamental structure of a DL algorithm is a neural network, comprised of interconnected nodes across multiple layers. This interconnected network enables the processing of intricate data in any given domain, allowing for tasks such as identifying patterns in large datasets, image processing, and natural language processing.

DL employs a statistical approach as it distributes a complex problem into many smaller modules and then employs ML algorithms on these modules, as shown in Figure 2.2. The neural networks are formed by a network of multiple connected layers. There are many interconnected networks of layers of nodes between the first layer, i.e., the input layer, and the last layer, i.e., the output layer. These layers in between are called hidden layers, which process a large amount of data to learn complex features of the data.

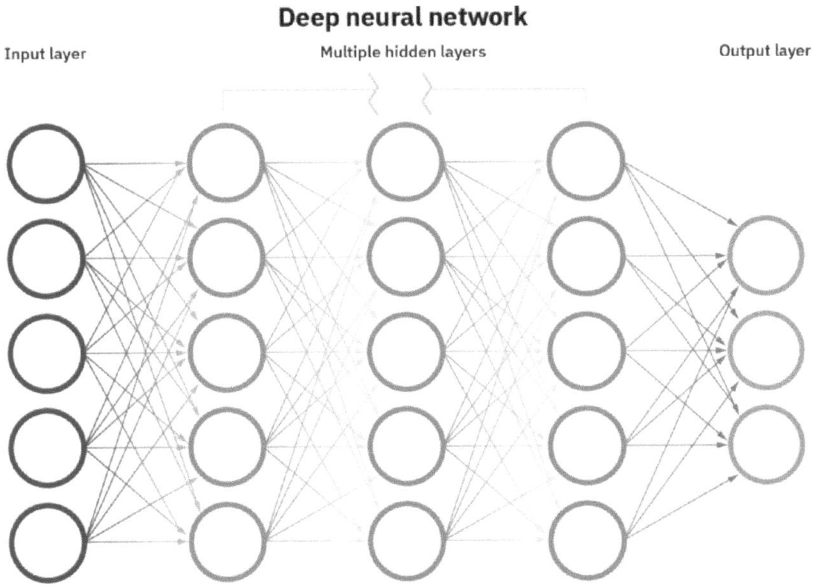

FIGURE 2.2 Architecture of neural networks.

Source: IBM.

Some of the common deep learning algorithms are:

- Auto encoders;
- Convolutional neural networks (CNNs);
- Deep belief networks (DBNs);
- Generative adversarial networks (GANs);

- Long short-term memory (LSTMs);
- Multilayer perceptron (MLPs);
- Radial basic function network (RBFNs);
- Recurrent neural networks (RNNs);
- Restricted Boltzmann machines (RBMs);
- Self-organizing maps (SOMs).

For training the machine in the research presented in this chapter, we used the CNN algorithm. CNNs, also known as ConvNets, are DL algorithms that take complex inputs, assign weights and biases at various levels, and derive valid and meaningful conclusions. While CNN algorithms are primarily used for image processing and analysis, their use in recommendation systems is also significant. Therefore, this research utilizes CNN to predict cyanotic CHD, employing a similar working method as a recommender system.

The architecture of the CNN algorithm suggests a multi-layered structure that processes and extracts features from data, as shown in Figure 2.3. Therefore, it can be utilized to train the machine with ideal cases and subsequently detect anomalies in testing cases.

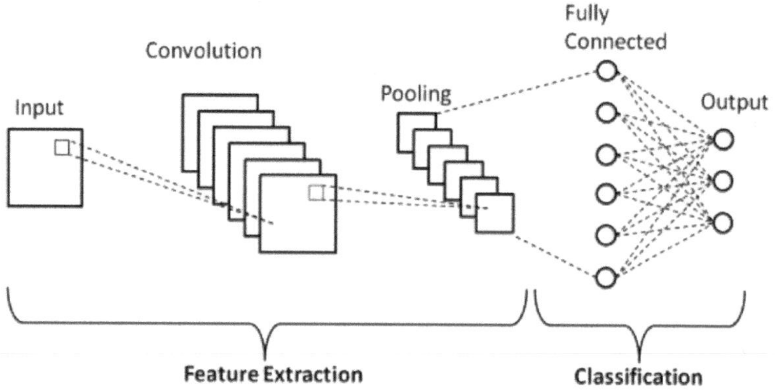

FIGURE 2.3 Architecture of CNN algorithm.

Source: Medium. https://medium.com/techiepedia/binary-image-classifier-cnn-using-tensorflow-a3f5d6746697

The CNN algorithm consists of following layers:

1. **Convolutional Layer:** This layer of CNN performs various convolution operations by employing several filters on the data.
2. **Rectified Linear Unit (ReLU):** This layer uses an activation function that triggers the next output function. The ReLU activation function

is a linear function that triggers the next output as 1 if it is positive, otherwise 0. The node employing the Rectified Linear activation function is called a ReLU. Neural networks using rectifier functions for hidden layers are called rectified networks. ReLU is employed because it provides more sensitivity to the activation sum input and avoids saturation. The output of this layer is a rectified feature map.

3. **Pooling Layer:** The output of the ReLU is passed on to the next layer, called the pooling layer, which reduces the feature map. This layer performs a down sampling operation. The final output of this layer is a single, long, continuous linear vector.
4. **Fully Connected Layer:** The final output is classified to identify anomalies and provide the ultimate verdict of the knowledge-based system, which is powered by the DL algorithm.

2.2 RESEARCH METHODOLOGY

Proposed algorithm as shown in Figure 2.4.

- **Step 1:** Device Configuration: First of all, a smart wearable like an Apple Watch that has the capacity to measure SpO_2 levels in the blood and collect the data in a nearby smartphone needs to be configured according to our needs.
- **Step 2:** Data Collection: A smart wearable can be used to collect the SpO_2 level of the children who are going to be tested for the likelihood of Cyanotic CHDs. The data for a certain period will be stored and prepared for processing and input into the neural network.
- **Step 3:** Decision Making: In this step, the normalized SpO_2 level of the children in a specific time interval will be analyzed. If it falls below the threshold level for being CHD free, an echocardiogram will be recommended.
- **Step 4:** CNN Algorithm's Verdict: The pattern exhibited in the SpO_2 level, along with the image analysis of echocardiogram using the recommender and image processing capacity of the CNN algorithm, will produce the final probabilities of occurrence of cyanotic CHD in children. These probabilities will then require a medical professional's conclusive decision.

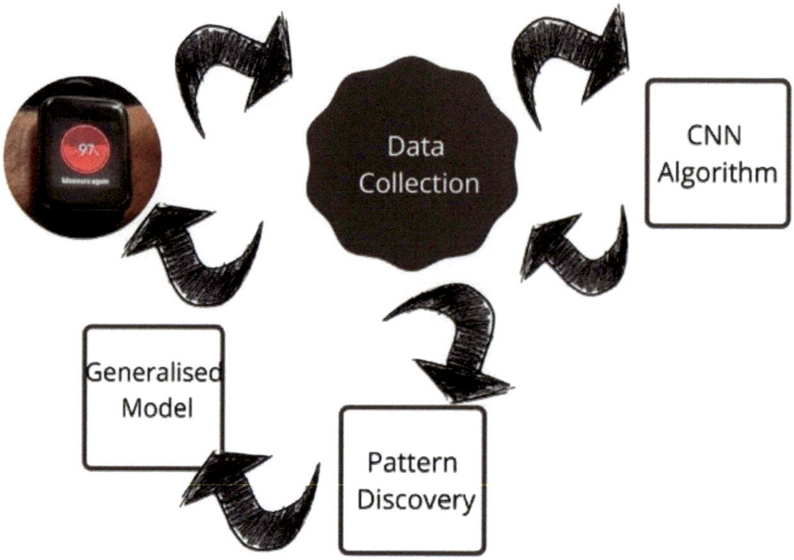

FIGURE 2.4 Proposed model.

2.3 LITERATURE REVIEW

A significant amount of work has previously been conducted in the field of medical sciences using AI. This is particularly evident in the area of cardiovascular diseases, where the potential role of AI in future medicines and treatment methodologies can be observed [3]. Similarly, CVDs are more prevalent in developing and underdeveloped regions such as Asia and Africa [4]. Data mining techniques have long been utilized to uncover healthcare mysteries. Physicians have directly benefited from the use of data mining tools to assess treatment effectiveness and determine best practices [5]. Additionally, knowledge discovery tools are employed to make decisions regarding various medical practices and disease diagnoses. In many cases, manual intervention alone is insufficient to arrive at a conclusive answer, which is where these knowledge discovery tools come into play [6]. Consequently, information sciences and AI have become primary tools in the field of medical sciences nowadays [7].

Not only DL and complex neural networks, but even the simplest of supervised learning methodologies are being used to solve issues in medical and healthcare systems, such as laboratory medicines [8]. DL algorithms, like CNNs, have been specifically used to analyze medical images. This image analysis has been particularly helpful in analyzing echocardiogram reports, which are the gold standard test for diagnosing cardiovascular diseases. Therefore, the CNN algorithm has great potential to address various unresolved issues in the diagnosis of cardiovascular diseases [9]. In this research, the use of CNN is approached in a similar way to that of a recommender system [10].

2.4 CONCLUSION AND FUTURE WORKS

Possible deaths and severities due to cardiovascular diseases can be entirely prevented if the diagnosis of these diseases is done at an early stage. Some of these diseases appear from birth, making it difficult to diagnose and predict their occurrence. Cyanotic congenital heart disease (CHD) is one such disease that this particular chapter is based on. The research methodology specifically aims to identify a pattern through which a generalized model can be developed to detect any occurrence of Cyanotic CHDs in children without the need for complex medical diagnostic methods. This research can also be further developed to diagnose some autoimmune and genetic cardiovascular diseases.

KEYWORDS

- **artificial intelligence**
- **cardiovascular diseases**
- **congenital heart disease**
- **convolutional neural networks**
- **deep learning**
- **long short-term memory**
- **rectified linear unit**
- **restricted Boltzmann machines**

REFERENCES

1. Kagawa, M., Takeshita, M., Yatō, S., & Kitamura, K., (1983). Brain abscess in congenital cyanotic heart disease. *Journal of Neurosurgery, 58*(6), 913–917.
2. Dhawan, S., Gupta, R., & Rana, A. K., (2021). Internet of things: Security issues, challenges, and its applications. In: *Multimodal Biometric Systems* (pp. 111–124). CRC Press.
3. Yan, Y., Zhang, J. W., Zang, G. Y., & Pu, J., (2019). The primary use of artificial intelligence in cardiovascular diseases: What kind of potential role does artificial intelligence play in future medicine? *J. Geriatr. Cardiol., 16*(8), 585–591. doi: 10.11909/j.issn.1671-5411.2019.08.010.
4. Hirotsugu, U., Akira, S., Katsuyuki, M., Tanvir, C. T., Naoyuki, T., Yoshikuni, K., Makoto, W., et al. (2021). *Cardiovascular Disease and Risk Factors in Asia.* doi: 10.1161/CIRCULATIONAHA.108.790048 https://doi.org/10.1161/CIRCULATIONAHA.108.790048.
5. Koh, H. C., & Tan, G., (2005). Data mining applications in healthcare. *J. Healthc. Inf. Manag., Spring, 19*(2), 64–72. PMID: 15869215.
6. Karys, A., (1998). Ending the technology paradox: Healthcare management technologies for clinical decision making. *J. Healthc. Inf. Manag. Fall, 12*(3), 25–32. PMID: 10338784.
7. Skinner, R. I., (2003). The value of information technology in healthcare. *Front Health Serv. Manage. Spring, 19*(3), 3–15. PMID: 12645778.
8. Hooman, H. R., Nam, T., Samer, A., & Luke, T. D. (2020). *Machine Learning in Health Care and Laboratory Medicine: General Overview of Supervised Learning and Auto-ML.* https://doi.org/10.1111/ijlh.13537.
9. Dormer, J. D., Halicek, M., Ma, L., Reilly, C. M., Schreibmann, E., & Fei, B., (2018). Convolutional neural networks for the detection of diseased hearts using CT images and left atrium patches. *Proc. SPIE Int. Soc. Opt. Eng., 10575*, 1057530. doi: 10.1117/12.2293548.
10. Dan, Y., Jing, Z., Sifeng, W., & Xue, D. Z., (2019). A time-aware CNN-based personalized recommender system. *Complexity, 2019*, 11. Article ID 9476981. https://doi.org/10.1155/2019/9476981.
11. Rana, A. K., & Sharma, S., (2019). Enhanced energy-efficient heterogeneous routing protocols in WSNs for IoT application. *Int J Eng Adv Technol (IJEAT), 9*(1).
12. Rana, A. K., Krishna, R., Dhwan, S., Sharma, S., & Gupta, R., (2019). Review on artificial intelligence with internet of things-problems, challenges, and opportunities. In: *2019 2nd International Conference on Power Energy, Environment, and Intelligent Control (PEEIC)* (pp. 383–387). IEEE.
13. Sharma, S., Dudeja, R. K., Aujla, G. S., Bali, R. S., & Kumar, N., (2020). DeTrAs: Deep learning-based healthcare framework for IoT-based assistance of Alzheimer patients. *Neural Computing and Applications*, 1–13.
14. Rajan, J. P., & Nadar, E. R. S., (2019). Smart-monitor: Patient monitoring system for IoT-based healthcare system using deep learning. *IETE Journal of Research*, 1–8.
15. Rana, A. K., & Sharma, S., (2021). Contiki Cooja security solution (CCSS) with IPv6 routing protocol for low-power and lossy networks (RPL) in internet of things

applications. In: *Mobile Radio Communications and 5G Networks* (pp. 251–259). Springer, Singapore.
16. Rana, A. K., & Sharma, S., (2021). Industry 4.0 manufacturing based on IoT, cloud computing, and big data: Manufacturing purpose scenario. In: *Advances in Communication and Computational Technology* (pp. 1109–1119). Springer, Singapore.
17. Rana, A. K., Sharma, S., Dhawan, S., & Tayal, S. (2021). Towards secure deployment on the internet of robotic things: Architecture, applications, and challenges. In: *Multimodal Biometric Systems* (pp. 135–148). CRC Press.
18. Kumar, A., & Sharma, S., (2021). Internet of things (IoT) with energy sector-challenges and development. In: *Electrical and Electronic Devices, Circuits and Materials* (pp. 183–196). CRC Press.
19. Kumar, A., & Sharma, S. (2021). IFTTT rely based a semantic web approach to simplifying Trigger-action programming for end-user application with IoT applications. *Semantic IoT: Theory and Applications: Interoperability, Provenance and Beyond*, 385.
20. Dhawan, S., Chakraborty, C., Frnda, J., Gupta, R., Rana, A. K., & Pani, S. K., (2021). SSII: Secured and high-quality steganography using Intelligent hybrid optimization algorithms for IoT. *IEEE Access*.
21. Paras, C., & Rajesh, K., (2014). Design and fabrication of a novel hexagonal reconfigurable antenna with RF switches for mobile terminal. *Journal of the National Science Foundation of Sri Lanka, 42*(3), 257–263. ISSN 1391-4588.
22. Paras, C., & Rajesh, K., (2015). A new spiral frequency reconfigurable antenna with RF- MEMS switches for mobile RF front end. *Applied Electromagnetics and Mechanics* (Vol. 47, pp. 323–335). IOS Press, Birmingham, Alabama 35201 USA, ISSN 1383–5416.
23. Paras, C., & Rajesh, K., (2016). A novel approach of design and analysis of fractal antenna using neuro-computational method for reconfigurable RF MEMS antenna. *Turkish Journal of Electrical Engineering & Computer Sciences, Academic Journal* (Vol. 24, No.3, pp. 1265–1278). Ankara – TURKEY (ISSN 1300-0632).
24. Anupma, G., Ankush, K., & Paras, C., (2019). Design of a patch antenna with square ring-shaped coupled ground for on-/off body communication. *International Journal of Electronics* (pp. 1814–1828). (TET N), Taylor & Francis, ISSN: 0020-7217.
25. Ayush, D., Bhawna, G., Paras, C., Apoorav, M. S., & Sanjeev, K., (2020). *Real-Time Image Dehazing Clarifies Digital Images* (Vol. 56, No. 1, pp. 91, 92). Laser Focus World, Pennwell Publ Co., USA, ISSN: 1043-809.
26. Dhawan, S., Gupta, R., & Rana, A. K., (2021). Steganography and steganalysis using machine learning. In: *Internet of Things* (pp. 249–257). CRC Press.

CHAPTER 3

AI AND IOT-BASED SMART HEALTHCARE WITH IMAGE PROCESSING

PRIYA TRIVEDI[1] and ARUN KUMAR[2]

[1]*Aisect College of Professional Studies, Indore, Madhya Pradesh, India*

[2]*Galgotias College of Engineering and Technology, Greater Noida, Uttar Pradesh, India*

ABSTRACT

Thanks to artificial intelligence (AI), people have made admirable strides in the field of healthcare. AI, along with other cutting-edge technology, has made healthcare facilities and procedures more effective and practical. Clinical and other applications that offer extremely effective treatment and perceptive operations are the consequence. One of the few technologies that has the ability to completely revolutionize healthcare systems throughout the world and assist in overcoming all industry difficulties is AI. AI may significantly reduce the amount of time it takes to offer healthcare services while also increasing effectiveness. A new high-technology healthcare system is now helping to discover and cure ailments all around the world. It is also extending our lives and boosting our communities' well-being. Health information is now secure, of greater quality, and easily accessible when and where it is needed, resulting in less expensive care and more efficiency for everyone through the adoption of secure health information technology. We now live in a new era of computing, in which all health-related

The Fusion of Artificial Intelligence and Soft Computing Techniques for Cybersecurity.
M. A. Jabbar, Sanju Tiwari, Subhendu Kumar Pani, & Stephen Huang (Eds.)

data is computerized. In the same context, our chapter focused on IoT and AI-based applications in healthcare systems (Health IT systems) that are utilized for disease diagnosis, prevention, and treatment of various diseases.

3.1 INTRODUCTION

We are not surprised by images. They are acquired by observing the objective world using various observing systems, each capable of directly or indirectly affecting the human eye and generating visual perception as a result. Human vision is an observational system, and the mental representation of the objective scene is the image it produces. Normal pictures are simulated images, meaning the information they contain is an analog quantity that is constantly changing. Therefore, the term "digital image processing" is often used to describe image processing on computers.

Retail, industrial, automotive, and now healthcare have all benefited from the internet of things (IoT) and AI. Although it is critical for all of these industries to incorporate these technologies to find answers, healthcare stands to benefit the most.

IoT has opened up new possibilities for doctors, nurses, patients, researchers, and even insurance. Doctors can now monitor their patients' health from almost anywhere, and patients can communicate with their doctors at their fingertips through the IoT. Researchers from around the world can collaborate without ever meeting, and professionals can get the help they need for even the most complex diagnostic problems.

Patients' interactions were limited to teleconferences, text messages, and visits only with doctors before the advent of the IoT. There was no means for doctors or hospitals to regularly assess patients' health and make appropriate recommendations.

In the healthcare industry, remote monitoring is now made possible through IoT-enabled gadgets/equipment, which can provide superior treatment for keeping patients healthy and safe by offering AI and IoT-designed clinical solutions. Further contacts with doctors are now more convenient and efficient. Additionally, it has also enhanced patient participation and satisfaction, while remote monitoring of a patient's health helps to prevent readmissions and shorten hospital stays.

IoT is clearly transforming the healthcare industry by redefining the role of technology and human touch in healthcare solution delivery. IoT applications in healthcare benefit patients, families, physicians, hospitals, and insurance companies.

The term "health information technology" refers to the computer systems used by healthcare professionals and patients to store, share, and analyze health data (health IT).

Examples of health information technology include:

1. **Electronic Medical Records (EMRs):** These enable doctors to keep better track of your health information and may allow them to see it even if their clinic is closed at the time, you have an issue. EHRs also make it simple for your doctor to communicate information with experts, ensuring that specialists who need your information get it as soon as possible.
2. **Personal Health Information (PHIs):** It is similar to an EHR, except that you have more control over the information that enters into it. A personal health record (PHR) can be used to keep track of information from doctor appointments, but it can also represent your life outside of the doctor's office and your health priorities, such as tracking what you eat, how often you exercise, and your blood pressure. Your PHI and your doctor's EHR may be linked at times.
3. **Prescriptions are Written Electronically (E-Prescribing):** A paper prescription can be misplaced or misinterpreted. Your doctor can connect directly with your pharmacy through e-prescribing. This means you can go to the drugstore without needing to have your paper prescription with you.
4. **Confidentiality and Safety:** All of these electronic solutions have the potential to improve the security of your health information. For instance, electronic data can be encrypted to ensure that only authorized individuals have access to it. Health information technology can also facilitate tracking of who has accessed your data.

3.2 LITERATURE SURVEY

Healthcare is an important aspect of everyone's life. With this in mind, IoT technology has received a lot of attention recently, and many researchers are investigating its potential for enhancing healthcare systems due to an aging population and an increase in chronic illnesses. By focusing on LPWANs and highlighting their specific suitability for use in IoT systems, it further contributes to the original research.

Existing IoT technologies are evaluated for their value in a Parkinson's disease patient monitoring system [1]. Several researchers have proposed practical methods for monitoring blood glucose levels in diabetic individuals. In this approach, patients manually measure their blood glucose levels at preset intervals [2]. A cellular phone equipped with electrocardiography (ECG) and blood pressure sensors transmits data on blood pressure, heart rate, body temperature, and oxygen saturation to a base station [3–5]. The system utilizes sensor data from accelerometers and magnetometers, as well as pedestrian dead reckoning, thresholding, and decision trees (DT), to determine a person's location within four-room zones and detect falls, lying down, standing, sitting, and walking actions. Some studies have presented fault-tolerance techniques for wireless body area network (BAN) healthcare monitoring systems [6]. The method is based on network coding, which employs a greedy grouping algorithm to establish a hierarchical network structure. Subsequently, linearly independent coding combinations are generated using random linear network coding [7]. In this chapter, the kidney exchange problem is modeled as a directed graph maximum-weight cycle-packing problem. According to the researchers, this approach may be used to allow one more patient to obtain a kidney transplant than current formulations' matching solutions. To enable secure communication between patients and caregivers over the internet, Rana & Sharma [8] propose a secure authentication and prescription safety strategy. The protocol is validated using Rubin logic formal modeling and NS-2 simulation. Zheng et al. [9] published a study on detecting protein complexes or function modules in protein-protein interaction networks. Their technique is based on the Gene Ontology and the simplified swarm optimization (SSO) algorithm. The SSO technique is utilized to cluster proteins with similar activities, while the GO is employed to enhance detection accuracy and aid in functional complex identification. The key principle is that the aortic component of the second heart sound waveform remains unaffected by breathing, whereas the pulmonary component exhibits a delay. To test this strategy, simulated second heart sounds with known changing splits were employed [10]. To design a unique statistical technique for discriminating ulcer from non-ulcer pixels, we utilized color bands as the feature vector and the support vector machine (SVM) with grid search as the classifier. In the laboratory, the approach was evaluated and found to have excellent accuracy, sensitivity, and specificity in diagnosing ulcers.

IoT can effectively connect people to people, things to things, and things to other things at any time and location. In an intelligent environment, sensors and gadgets invisibly connect over the IoT, making cross-platform information exchange simple. Digital photos are a significant component of the information in the IoT, which can connect various types of information. The link between image processing and IoT is depicted in Figure 3.1. The image sensor, made of CCD or CMOS and used to capture image data, is a component of the IoT network.

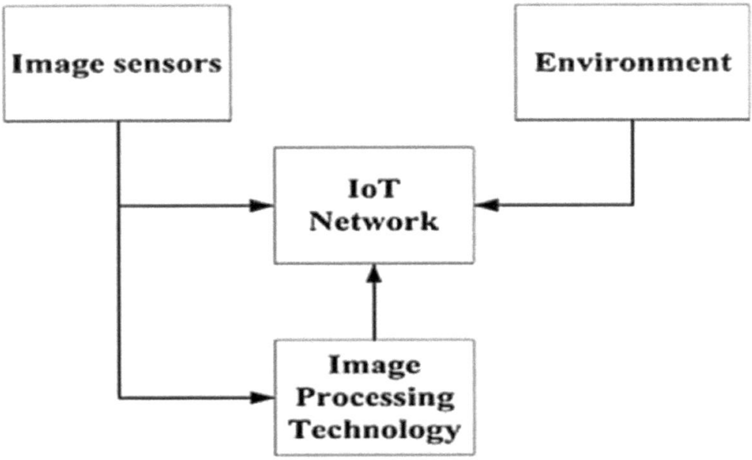

FIGURE 3.1 Relationship between IoT and image processing.

Source: The Rise of Internet of Things and Smart Cities. https://ebrary.net/201930/engineering/internet_things_image_processing

The image sensor serves as the analog of a person in the IoT network and is used to collect modified pictures from the external world. The visual processing region of the human brain, which is utilized to analyze and enhance these images, is analogous to the processor equipped with DIP technology. The processor then transmits the transformed result to other IoT network nodes so that they can respond to the task. Consequently, the IoT network equipped with DIP technology can obtain more accurate information and respond to tasks more quickly.

3.3 IMPROVING INFORMATION IN HEALTHCARE SYSTEMS

Significant efforts have been undertaken in recent years to reform health systems as a whole, with a greater emphasis on decentralizing resources and decision-making to the district level. Such transitions will necessitate improved health information systems at the local level. At the same time, it is clear that local people and families must be involved in the collection, dissemination, and analysis of health data. IoT devices present a host of new opportunities for healthcare practitioners to track patients as well as their patients' progress. As a result, the numerous wearable IoT devices bring both benefits and issues for both patients and healthcare practitioners.

3.3.1 REMOTE PATIENT MONITORING (RPM)

In order to investigate the application of image processing technology in remote monitoring and intelligent medical systems, the idea and implementation technique of a remote intelligent image processing system based on a virtual local area network are presented. This technique examines the essential technologies to be addressed in remote picture monitoring and employs advanced digital image compression coding and decoding technology, as well as digital image transmission technology. Remote patient monitoring (RPM) is the most popular use of IoT devices in healthcare. Patients who are unable to travel to a hospital, clinic, or healthcare facility can utilize IoT devices to gather health indicators and data such as blood pressure, temperature, and heart rate [11–16].

First, information about the patient is collected by an IoT device, which then sends it to a software application where healthcare professionals and/or patients can easily view it. Next, algorithms are used to evaluate the data and provide treatment suggestions or recommendations. For example, if the patient's IoT sensor detects a low heart rate, it triggers an alarm so that healthcare practitioners can intervene. An RPM system is shown in Figure 3.2. One of the key challenges with RPM devices is ensuring the security and privacy of the highly personal data collected by IoT devices [17–22].

In some manner, all wearables must convey patient data to their healthcare providers, which is commonly done through a specialized smartphone app. While there are a variety of protocols that may be used to communicate data from RPM devices to smartphones, the most common is Bluetooth low energy (BLE) communication. Figure 3.3 shows types of RPM devices.

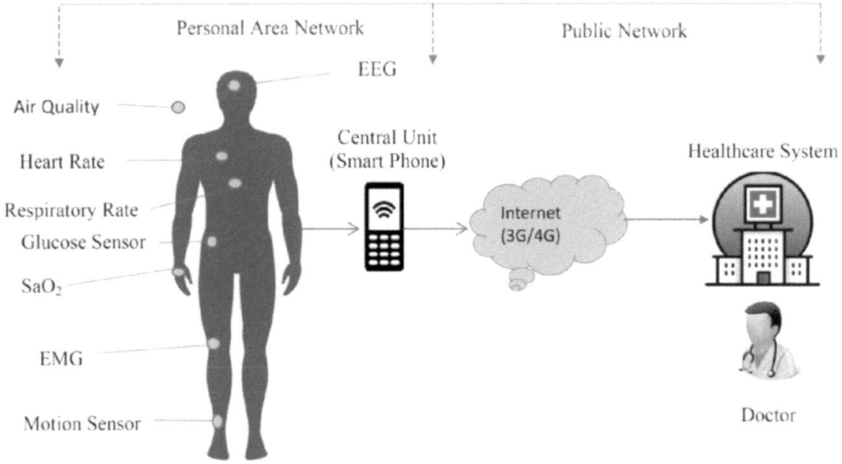

FIGURE 3.2 Remote patient monitoring system with image processing.

Source: Adapted from Kumar et al., 2022.

3.3.2 GLUCOSE MONITORING

Glucose monitoring has long been challenging for millions of Indians, as well as Americans with Type 1 or Type 2 diabetes. Not only is manually checking and documenting glucose levels burdensome, but it only provides information about a patient's glucose levels at the time of the test. If levels fluctuate significantly, periodic testing may not be sufficient to identify a problem.

IoT-enabled devices can help with these difficulties by providing continuous and automated glucose monitoring in patients. Glucose monitoring devices reduce the need for human record-keeping by alerting patients when their glucose levels are dangerously low or high.

Designing glucose monitoring devices through IoT should be:

- Small, easy to carry, and continuously monitor the patient's glucose level;
- Consuming less electricity than it needs to be recharged regularly is one of the challenges.

Figure 3.4 showing a glucose monitoring device that faces numerous obstacles, but the technology that addresses them has the potential to transform the way patients monitor their glucose levels.

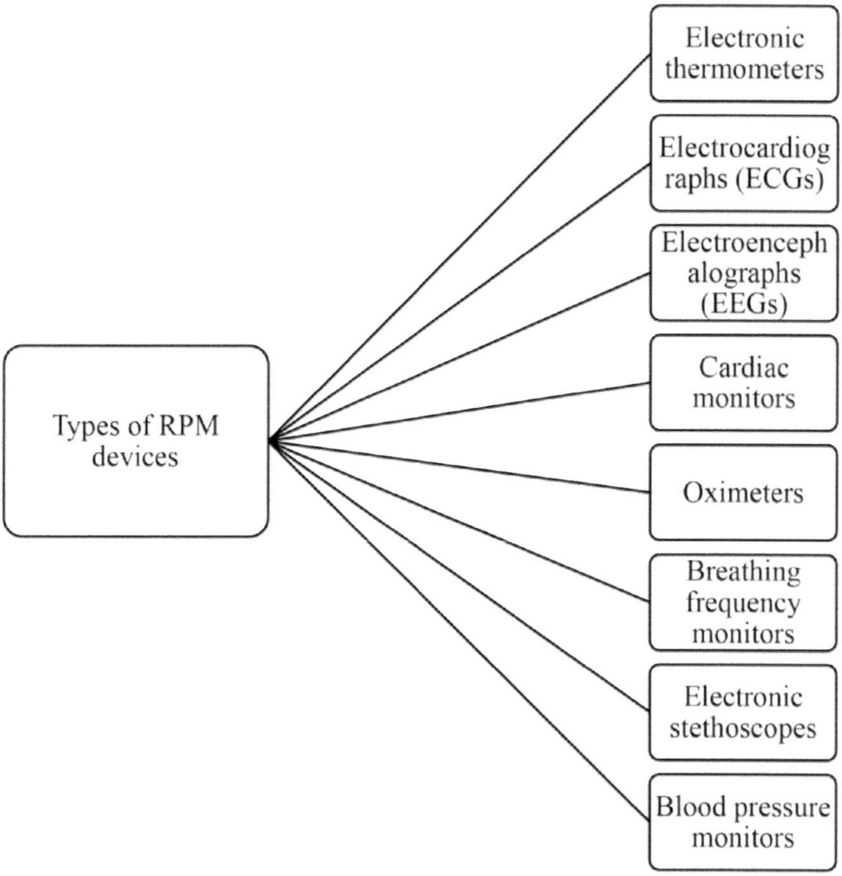

FIGURE 3.3 Types of RPM devices.

3.3.3 MONITORING OF HEART RATE

In healthcare facilities, even for patients, constant monitoring of heart rates, glucose levels, and other vital signs can be challenging. Periodic heart rate monitoring cannot protect against rapid heart rate changes. Even traditional hospital equipment used for continuous monitoring of cardiac activities makes patients permanently tied to wired machines, limiting their movement.

Today, there are various small IoT devices available for heart rate monitoring, allowing patients to freely move around while their hearts are constantly monitored. Although achieving ultra-accurate readings remains

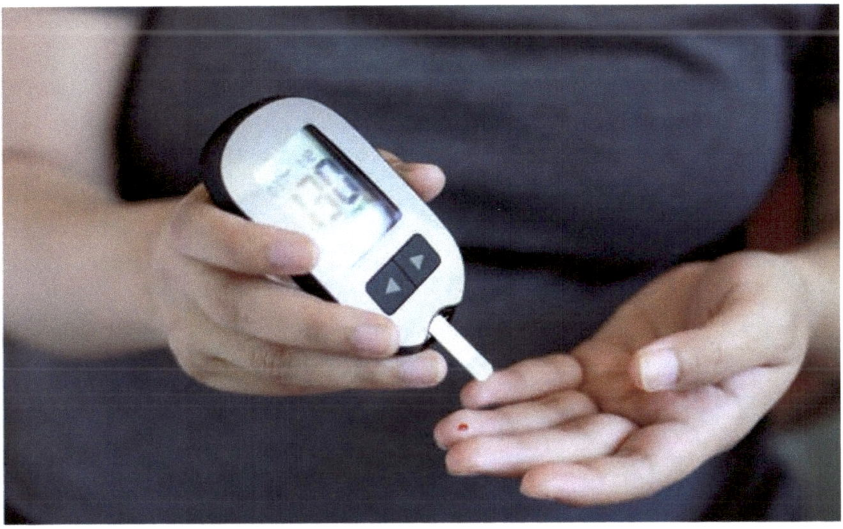

FIGURE 3.4 Glucose monitoring device.

Image credit: Kwangmoozaa (iStock)

a challenge, most modern equipment can achieve accuracy rates of 90% or higher. Measuring heart rates using IoT is shown in Figure 3.5.

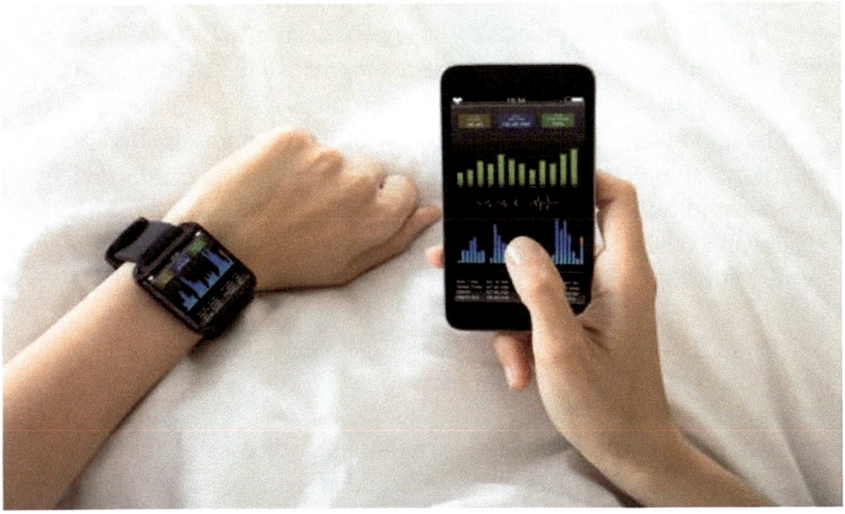

FIGURE 3.5 Measuring heart rates through IoT.

Image credit: AndreyPopov (iStock)

3.3.4 HAND HYGIENE MONITORING

In the past, there was no way to guarantee that physicians and patients in a healthcare institution thoroughly washed their hands to limit the risk of spreading infection. IoT devices are now being used in many healthcare institutions and hospitals to remind patients to wash their hands before entering the hospital and their rooms. The new design IoT devices can also provide instructions on how to clean your hands to decrease a specific risk for a particular patient. A major issue with these gadgets is that they can only remind individuals to wipe their hands, not do it for them. Nonetheless, data shows that these new design IoT devices can reduce hospital infection rates by more than 60%.

3.3.5 DEPRESSION AND MOOD MONITORING

Depressive symptoms information of patients and data on overall mood have traditionally been difficult to gather regularly. Patients could inform healthcare providers about their feelings on a regular basis, but they couldn't anticipate mood fluctuations. Patients often struggle to express their emotions appropriately. IoT devices that are "mood-aware" can help address these concerns. By collecting and analyzing data such as heart rate, mood, and blood pressure, these new design IoT devices can gather or predict information about a patient's mental state. Advanced IoT mood monitoring devices may even track data such as a patient's eye movement.

The main challenge is that this type of data cannot effectively predict depressive symptoms or other warning signs. While wearable devices like the ones mentioned above are the most common form of IoT device in healthcare, other gadgets go beyond monitoring and provide therapy or even reside "live" in or on the patient.

3.3.6 CONNECTED INHALERS

Asthma and chronic obstructive pulmonary disease (COPD) are two illnesses that commonly result in sudden and unprovoked episodes. Patients can benefit from IoT-connected inhalers since they can track the frequency of attacks and collect data from the environment to assist healthcare professionals in determining what prompted an attack. Furthermore, connected inhalers can notify patients when they leave their inhalers at home, putting them at risk of suffering an attack without it, or when they misuse their inhalers.

3.3.7 CONTACT LENSES WITH AN INTERNET CONNECTION

Smart contact lenses provide yet another method for passively and non-intrusively collecting health data. They may also include micro-cameras that allow wearers to take pictures with their eyes, which is why Google has patented connected contact lenses.

Smart glasses have the potential to turn human eyes into a powerful tool for digital interactions, whether used to enhance health outcomes or for other purposes.

3.3.8 ROBOTIC SURGERY

By implanting microscopic, Internet-connected robots into the human body, surgeons can perform complex procedures that would be difficult to accomplish with human hands. Simultaneously, robotic operations using small IoT devices can minimize the size of incisions required for surgery, resulting in a less invasive process and faster patient recovery.

These devices must be tiny and reliable enough to minimize disruption during surgery. They must also be capable of interpreting complex internal conditions to make the best possible decisions during surgery. However, IoT robots are currently being used in surgery, demonstrating that these challenges can be effectively addressed.

3.4 ADVANTAGES OF USING IOT AND IMAGE PROCESSING TECHNOLOGY IN HEALTHCARE

3.4.1 PATIENT-CENTRIC CARE

Medical IoT enables patients to become active partners in their treatment, resulting in better care. In addition, IoMT improves patient satisfaction with overall care and facilitates proper communication with medical personnel, enhancing physical comfort through features like climate control and smart lighting.

3.4.2 CARE-RELATED COSTS ARE REDUCED

Many patients who are not in immediate danger can utilize cloud-connected medical IoT devices from the comfort of their homes. They can track, collect, and communicate their health data, and continue receiving medical care remotely. Additionally, patients have the option to schedule e-visits with nurses and doctors through telehealth technology, eliminating the need to physically visit a hospital.

3.4.3 REDUCTIONS IN READMISSIONS

Patient apps with biosensors help ensure adherence to a discharge plan, provide early identification of health state changes, and allow for remote communication with a healthcare professional.

3.5 FUTURE PERSPECTIVES

In this section, we will examine the future of IoT in the healthcare sector. The IoT has proven to be an invaluable tool in the healthcare industry, providing advanced healthcare services to patients, clinicians, and researchers. Services such as smart diagnosis, wearable health tracking devices, patient management, and many others are available. Furthermore, IoT devices have helped alleviate the strain on the healthcare system [23].

Over a secure network, healthcare gadgets can deliver patients' health data directly to doctors. This enables clinicians to diagnose patients in remote areas:

1. **Healing at Home:** With the advancement of technology and the development of IoT health monitoring systems, healing at home will become a viable alternative. Patients will no longer need to be treated in a hospital bed through the integration of real-time monitoring and other components.
2. **Mindfulness:** The integration of wearable technology makes healing at home a possibility, not just potential. Wearable devices can track a patient's condition while also informing their doctor if something is wrong.

3. **Independent Health Surveillance:** You've probably heard about or had personal experience with independent health monitoring. It may have been if you used a smart watch to track your heart rate or a smartphone app to monitor how you slept, but with IoT progressing at such a quick pace, these incredible innovations will become obsolete in no time [24–27].

   The IoT is predicted to promote independent health monitoring, allowing patients to be assisted from anywhere in the world. If you have a health problem or are injured and unable to call for help, the health monitoring system will be able to respond and automatically notify your doctors or caregivers. IoT has a lot of potential in terms of assisting retirees and disabled people.

4. **Medicine at the Right Time:** How many times have you failed to remember to take your medication on time? The majority of us have. It appears that forgetting is in our DNA as humans. Patients may feel certain that they are taking the proper medicine at the right time through the use of IoT monitoring devices. Devices like these could assist patients with dementia or Alzheimer's disease to keep track and reduce confusion for both patients and caregivers.

The IoT is transforming the healthcare industry. There may be a few glitches to iron out over time, but the possibilities of IoT in healthcare are limitless. IoT will not only assist in simplifying the administrative side of healthcare, but it will also improve patient care and save lives [28].

3.6 CONCLUSION

The chapter defines the major distributor of healthcare systems as one of the most important applications of IoT. It removes geographic, temporal, and other limitations, allowing people to access healthcare at any time and in any location, while also improving coverage and efficiency. Thanks to the IoT health revolution, consumers can now receive high-quality care at an affordable cost. These applications generate a large amount of sensor data, which needs to be properly monitored and managed. In the health sector, cloud computing is a viable option for efficient knowledge processing. The provided framework is unique and can be utilized in various ways. It enables the management of patient-specific data from cloud devices and networks. The cloud app, built on IoT and its design principles, facilitates direct communication between sensor devices and serves stored data, users, and sensors. Wireless sensor networking provides a single point of access to embedded sensor control systems and the entire system service. The goal of this chapter is to establish a fully integrated IoT-based healthcare system that recognizes the need for better integration of various IoT services. Safety issues still need to be addressed in relation to the different implementation phases.

KEYWORDS

- **artificial intelligence**
- **cloud computing**
- **health IT system**
- **healthcare systems**
- **internet of things**
- **remote patient monitoring**
- **sensor control systems**

REFERENCES

1. Rana, A. K., & Sharma, S., (2021). The fusion of blockchain and IoT technologies with Industry 4.0. In: *Intelligent Communication and Automation Systems* (pp. 275–290). CRC Press.
2. Ali, F., El-Sappagh, S., Islam, S. M. R., Kwak, D., Ali, A., Imran, M., & Kwak, K. S., (2020). A smart healthcare monitoring system for heart disease prediction based on ensemble deep learning and feature fusion. *Inf. Fusion, 63*, 208–222.
3. Kumar, A., & Sharad, S., (2021). IFTTT rely based a semantic web approach to simplifying trigger-action programming for end-user application with IoT applications. In: *Semantic IoT: Theory and Applications* (pp. 385–397). Springer, Cham.
4. Kaur, S., Singla, J., Nkenyereye, L., Jha, S., Prashar, D., Joshi, G. P., El-Sappagh, S., et al., (2020). Medical diagnostic systems using artificial intelligence (AI) algorithms: Principles and perspectives. *IEEE Access, 8*, 228049–228069. [Google Scholar] [CrossRef].

5. Afzal, M., Islam, S. M. R., Hussain, M., & Lee, S., (2020). Precision medicine informatics: Principles, prospects, and challenges. *IEEE Access, 8*, 13593–13612. [Google Scholar] [CrossRef].
6. Rana, A. K., Ram, K., Sachin, D., Sharad, S., & Rashmi, G., (2019). Review on artificial intelligence with internet of things-problems, challenges, and opportunities. In: *2019 2nd International Conference on Power Energy, Environment, and Intelligent Control (PEEIC)* (pp. 383–387). IEEE.
7. Kumar, A., Sharma, S., Goyal, N., Singh, A., Cheng, X., & Singh, P., (2021). Secure and energy-efficient smart building architecture with emerging technology IoT. *Comput. Commun., 176*, 207–217. [CrossRef].
8. Rana, A. K., & Sharma, S., (2019). Enhanced energy-efficient heterogeneous routing protocols in WSNs for IoT application. *Int. J. Eng. Adv. Technol., 9*, 4418–44250.
9. Pasluosta, C. F., Gassner, H., Winkler, J., Klucken, J., & Eskofier, B. M., (2015). An emerging era in the management of Parkinson's disease: Wearable technologies and the internet of things. *IEEE Journal of Biomedical and Health Informatics, 19*(6), 1873–1881.
10. Chang, S. H., Chiang, R. D., Wu, S. J., & Chang, W. T., (2016). A context-aware, interactive M-health system for diabetics. *IT Professional, 18*(3), 14–22.
11. Chung, W. Y., Lee, S. C., & Toh, S. H., (2008). WSN-based mobile u-healthcare system with ECG, blood pressure measurement function. In: *Proceedings of the 2008 30th Annual International Conference of the IEEE Engineering in Medicine and Biology Society* (pp. 1533–1536). Vancouver, BC, Canada.
12. Tan, S. L., Garcia-Guzman, J., & Villa-Lopez, F., (2012). A wireless body area network for pervasive health monitoring within smart environments. In: *Proceedings of the 2012 IEEE International Conference on Consumer Electronics (ICCE-Berlin)* (pp. 47–51). Berlin, Germany.
13. Dziak, D., Jachimczyk, B., & Kulesza, W. J., (2017). IoT-based information system for healthcare application: Design methodology approach. *Appl. Sci., 7*, 596.
14. Peng, Y., Wang, X., Guo, L., Wang, Y., & Deng, Q., (2017). An efficient network coding-based fault-tolerant mechanism in WBAN for smart healthcare monitoring systems. *Appl. Sci., 7*, 817.
15. Yuh, J., Chung, S., & Cheong, T., (2017). Reformulation-linearization technique approach for kidney exchange program IT healthcare platforms. *Appl. Sci., 7*, 847.
16. Mahmood, Z., Ning, H., Ullah, A., & Yao, X., (2017). Secure authentication and prescription safety protocol for telecare health services using ubiquitous IoT. *Appl. Sci., 7*, 1069.
17. Zheng, X., Wu, L., Ye, S., & Chen, R., (2017). Simplified swarm optimization-based function module detection in protein–protein interaction networks. *Appl. Sci., 7*, 412.
18. Tang, H., Chen, H., & Li, T., (2017). Discrimination of aortic and pulmonary components from the second heart sound using respiratory modulation and measurement of respiratory split. *Appl. Sci., 7*, 690.
19. Suman, S., Hussin, F. A., Malik, A. S., Ho, S. H., Hilmi, I., Leow, A. H. R., & Goh, K. L., (2017). Feature selection and classification of ulcerated lesions using statistical analysis for WCE images. *Appl. Sci., 7*, 1097.
20. Rghioui, A., Lloret, J., Harane, M., & Oumnad, A., (2020). A smart glucose monitoring system for diabetic patient. *Electronics, 9*, 678. [Google Scholar] [CrossRef].

21. Dhawan, S., Chakraborty, C., Frnda, J., Gupta, R., Rana, A. K., & Pani, S. K., (2021). SSII: Secured and high-quality steganography using Intelligent hybrid optimization algorithms for IoT. *IEEE Access*.
22. Paras, C., & Rajesh, K., (2014). Design and fabrication of a novel hexagonal reconfigurable antenna with RF switches for mobile terminal. *Journal of the National Science Foundation of Sri Lanka* (Vol. 42, No. 3, pp. 257–263). ISSN 1391-4588.
23. Paras, C., & Rajesh, K., (2015). A new spiral frequency reconfigurable antenna with RF- MEMS switches for mobile RF front end. *Applied Electromagnetics and Mechanics* (Vol. 47, pp. 323–335). IOS press, Birmingham, Alabama 35201 USA, ISSN 1383-5416.
24. Paras, C., & Rajesh, K., (2016). A novel approach of design and analysis of fractal antenna using neuro-computational method for reconfigurable RF MEMS antenna. *Turkish Journal of Electrical Engineering & Computer Sciences, Academic Journal* (Vol. 24, No.3, pp. 1265–1278). Ankara – TURKEY, ISSN 1300-0632.
25. Rana, A. K., & Sharma, S., (2021). Contiki cooja security solution (CCSS) with IPv6 routing protocol for low-power and lossy networks (RPL) in Internet of Things Applications. In: *Mobile Radio Communications and 5G Networks* (pp. 251–259). Springer: Singapore.
26. Anupma, G., Ankush, K., & Paras, C., (2019). Design of a patch antenna with square ring-shaped coupled ground for on-/off body communication. *International Journal of Electronics* (pp. 1814–1828). (TET N), Taylor & Francis, ISSN: 0020-7217.
27. Ayush, D., Bhawna, G., Paras, C., Apoorav, M. S., & Sanjeev, K., (2020). *Real-Time Image Dehazing Clarifies Digital Images* (Vol. 56, No. 1, pp. 91, 92). Laser Focus World, Pennwell Publ Co., USA, ISSN: 1043-8092.
28. Rana, A. K., & Sharma, S., (2021). Industry 4.0 manufacturing based on IoT, cloud computing, and big data: Manufacturing purpose scenario. In: *Advances in Communication and Computational Technology* (pp. 1109–1119). Springer: Singapore.
29. Kumar A, Sharma S, Singh A, Alwadain A, Choi B-J, Manual-Brenosa J, Ortega-Mansilla A, Goyal N. Revolutionary Strategies Analysis and Proposed System for Future Infrastructure in Internet of Things. Sustainability. 2022; 14(1):71. https://doi.org/10.3390/su14010071

CHAPTER 4

CYBER PHYSICAL SYSTEM IN MEDICAL SCIENCE

NITIN TYAGI, POONAM RANI, SIDDHARTH GAUTAM, and PRIYANSHU JOSHI

Department of Computer Science and Engineering, GGSIPU, Delhi, India

ABSTRACT

Medical cyber physical systems (MCPS) are critical-to-life systems with the priority of ensuring patient safety. They guarantee the validation of technologies and propose a safer and more efficient approach to healthcare, aiming to increase productivity without compromising treatment effectiveness. The basic approach is to utilize existing technologies in an organized and advanced manner, expanding their reach. MCPS is a part of the cyber physical system (CPS) trend, which is already solving numerous problems worldwide. MCPS includes patient monitoring through various devices and introduces newer technologies. It prioritizes safety and transparency in work, addressing previously overlooked or easily managed issues. This chapter presents a comprehensive overview of MCPS and CPS and explores how their implementation will benefit the modern world.

4.1 INTRODUCTION

The cyber physical systems (CPS) indicate the new generation of systems that have unified capabilities and functionalities of interacting with

The Fusion of Artificial Intelligence and Soft Computing Techniques for Cybersecurity.
M. A. Jabbar, Sanju Tiwari, Subhendu Kumar Pani, & Stephen Huang (Eds.)

humans and bringing closeness to the virtual world with the real world. Over the years, research has been carried out to find ways to ease the tasks that humans find exhausting and time-consuming [1]. CPS not only reduce the efforts of humans but also provide extra functionality with the devices and greater efficiency. Industries and companies have developed various systems that best suit the situation in which the work extraction is doubled while the workload remains the same [2]. The increasing complexity of advanced systems and technologies challenges us to get the best out of them and prepare along with the best efforts possible. The need to build next-gen control systems has now become a necessity. The CPS need a structured model of tools and mechanisms to fulfil the needs of the manufacturer and the consumer. Computer Science has made enough advancements in the field of modern technology and in the upcoming times, we are expecting it to diversify [3]. The CPS will play a major role in various domains that will help mankind achieve heights that were not possible a few years back [4]. We need the right theories and tools to implement the CPS accurately and help us get cost-effective, reliable, and efficient resources so as to resolve the problems easily and at a quicker speed [5]. The basic steps to be taken care of while developing a CPS include designing, analyzing, and verifying the used components, system, and software architecture, conducting regular stage checks to monitor performance, efficiency, accuracy, and areas of improvement. CPS research is still in its early years, with isolated attention given to sensor devices, communication and networking devices, control theories, mathematical calculations, software engineering, and computer science. By focusing on these areas, we can extract the best machinery ever produced, which will help improve current standards of living, healthcare, and modern requirements that affect our daily lives. The advancement in the field of CPS can be accelerated by identifying opportunities, challenges, and needs in various sectors of industrial development [6]. The objective is to build new technologies that are compatible, energy-efficient, low maintenance, and most importantly, efficient. The security of the user must also be a concern, and the device or machinery must comply with the rules of development.

Cyber physical systems (CPS) in the medical field refer to a critical network of healthcare services. Instead of standalone devices currently used in healthcare, which monitor or suggest treatment for individuals, CPS will soon be enhanced to monitor multiple patients and assist healthcare workers simultaneously. This will involve a coherent display of patients' health conditions, joining the virtual world with the real world. Data, medicine, prescriptions, and basic details will be easily and securely accessible only to authorized personnel [7]. This revolutionary trend has greatly impacted the CPS sector.

The development of a safe and secure medical cyber physical system (MCPS) requires perfect design, architecture, structured planning, and validation and verification techniques. Due to the increased complexity and size of this sector, model-based technology is necessary. MCPS designs also encompass communication and devices within the entire scenario. MCPS will open up new opportunities in the healthcare sector, not only in terms of advancement but also cost-effectiveness. However, obtaining approval for the use of MCPS in patient treatment requires a new regulatory and managed procedure. The current process-based regulatory regime used by the U.S. Food and Drug Administration (FDA) is quite extensive, costly, and complex. Figure 4.1 illustrates the workflow of a medical organization and depicts the actual process of analyzing the credibility of an individual.

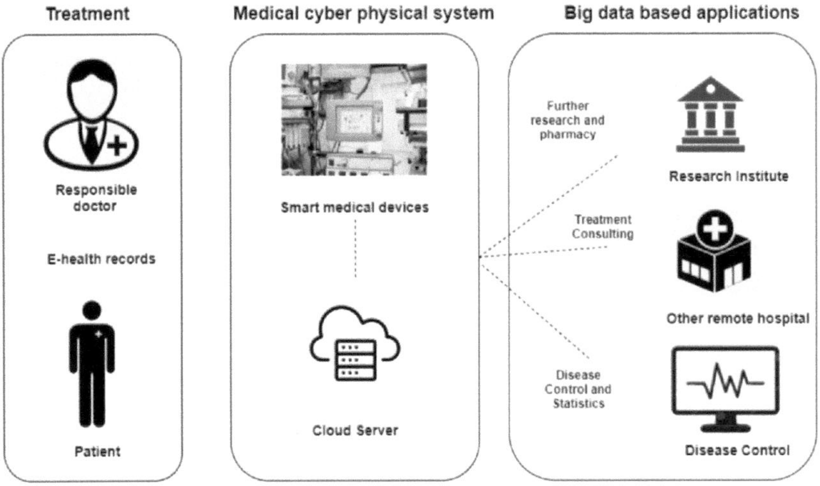

FIGURE 4.1 Medical cyber physical system workflow.

The remaining chapter is organized as follows: Section 4.2 discusses the taxonomy of MCPS. In Section 4.3, the devices currently in use will be discussed. Section 4.4 will include the challenges faced by the MCPS systems and how they can be solved, as well as the extent to which we can extract the best out of them. Section 4.5 includes the advantages of MCPS that will help improve our healthcare sector, and Section 4.6 will contain the conclusion along with the future scope of MCPS.

4.2 CATEGORIZATION OF CPS IN HEALTH SECTOR

The devices/systems displayed above are discussed in detail below. These systems contribute to the improvement of MCPS as a whole and suggest the ongoing technologies in the market, along with the changes that can be made to enhance them. Figure 4.2 depicts the taxonomy of CPS in healthcare, which includes six systems or devices.

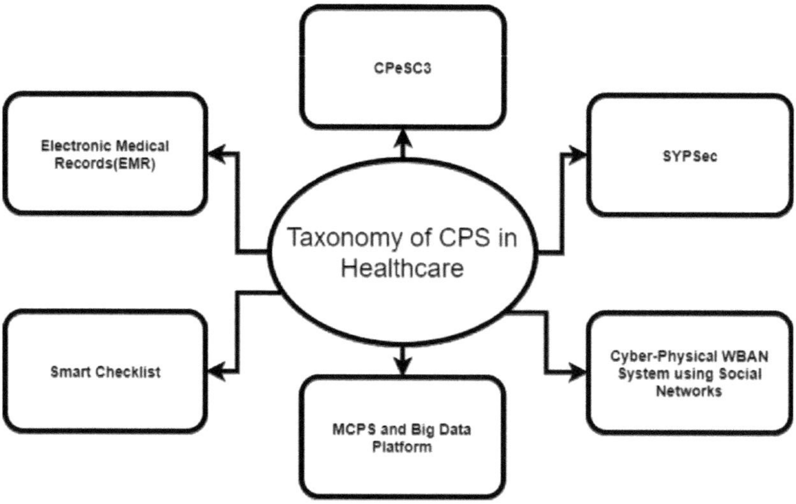

FIGURE 4.2 Taxonomy of CPS in healthcare.

4.2.1 ELECTRONIC MEDICAL RECORDS (EMR)

Electronic medical records (EMR) refer to the structure of a virtual-physical interface used for automatic readings of vital signs. This interface serves as a solution to the time-consuming and error-prone process of monitoring. The design of the interface allows the network to collect and store data into the EMR system. The key elements of this design include the EMR, a software adapter known as the data handler, and a vital sign reading station based on hardware topology [8]. The system follows a three-tier architecture, as discussed in the SDD, to address the virtual-physical approach. This approach ensures that changes made to components in one layer do not affect the components in other layers [9]. The implementation of EMR offers several benefits. Patient data can be tracked over a specific timeline, and it can be accessed by authorized healthcare contributors. This facilitates the identification of patients who are due for routine check-ups and appointments, as well as monitoring their specific needs such as vaccinations and blood pressure readings. EMRs are designed to support organizations in providing systematic and accurate care to patients.

4.2.2 CPeSC3

Proposed by Wang et al. the layout of CPeSC3 consists of three major components: Communication Core, Computation Core, Resource Scheduling, and Management Core. Cloud computing technology is adopted, specifically in reference to healthcare applications. A healthcare app scenario is presented here, which is established on a practical test bed for confirmation purposes [10]. The advantages of CPeSC3 include increased functionality and validation of treatments provided by physicians. It enables the healthcare system to easily monitor large amounts of data, reduce redundancy, and provide efficient service in a timely manner.

4.2.3 CYPSec

Proposed by Venkatasubramanian et al. cyber physical security (CYPSec) is a solution that is synchronized with other security solutions, incorporating traditional security measures and environmental aspects [11]. The design principles and challenges of CYPSec can be illustrated through two examples: Physiological signal-based key agreement (PSKA) and critically aware access control (CAAC). PSKA facilitates automatic key agreement between sensor devices in the body area network (BAN) based on physiological signals from the body. On the other hand, CAAC provides authority to establish and manage emergency situations responsibly. CYPSec offers several advantages. It combines traditional methodologies with modern methods, ensuring the security of patient data and fostering trust in organizations. The data is not easily manipulated, guaranteeing accurate and effective healthcare services.

4.2.4 VIRTUAL-PHYSICAL WBAN SYSTEM THROUGH SOCIAL NETWORK

Zhang et al. propose an approach that utilizes power games to mitigate communication disruptions in WBANs. This approach is based on the information that encompasses social interactions among individuals. The contribution is made in the following areas: structuring inter-WBAN disturbances and calculating the distance distribution of interruptions, creating algorithms for detecting and predicting social communications for individuals carrying WBANs [12]. The Cyber-Physical WBAN System using Social Networks has several advantages: it enhances interaction, reduces the distance between patients and doctors, allows healthcare service providers to closely engage with patients' families, and provides better guidance for the future. Technological advancements have streamlined the interaction between patients and doctors, making it easier [13].

4.2.5 MCPS AND BIG DATA PLATFORM

A Big Data processing infrastructure for MCPS is proposed by Don & Dugki, which integrates the physical world with dynamics and provides flexibility to the virtual world in healthcare decision-making systems. This framework claims to reduce hospital expenses and simplify routine check-ups conducted by clinic doctors. Additionally, it will help enhance facilities in remote areas by transmitting data collected from ECG machines, oxygen levels, and pulse rates through various sensors to a remote application server. Physicians can then analyze this data to monitor the condition of all patients and prescribe accordingly [14]. Benefits: Big Data in MCPS establishes a fully aware system, ensuring efficient and trouble-free delivery. This system requires various subsystems to perform tasks such as filtering, listening, processing, accelerating, and enrichment, ultimately creating a fully equipped system and enhancing overall functionality [15]. Big Data is a growing trend, and with the right resources and guidance, it can flourish for the betterment of the healthcare sector and improve delivery.

4.2.6 SMART CHECKLIST

A Smart Checklist is proposed by Avrunin et al. to support and assist participants with their respective tasks. This system also facilitates the interaction between devices and software applications. When used in intensive care, this system is intended to aid medical professionals in medication administration, data collection, and other standard activities for patients. However, the system is not yet applicable to all medical procedures and requires further modification [16]. The advantages of Smart Checklist include improved assistance, faster responses, and more efficient results. Smart Checklist helps enhance services and reduces the likelihood of errors, which is another benefit of using it.

4.3 DEVICES USED IN MCPS

The devices used in healthcare not only ease the doctor's and staff's work but also provide convenience to the patients. The devices available in the market enable the patients to track their health, previous records, appointments, and other medical-related stuff online. This has revolutionized the whole system and has given a better approach for the same [17]. The devices used in MCPS are listed in subsections. Figure 4.3 shows the devices in a systematic behavior.

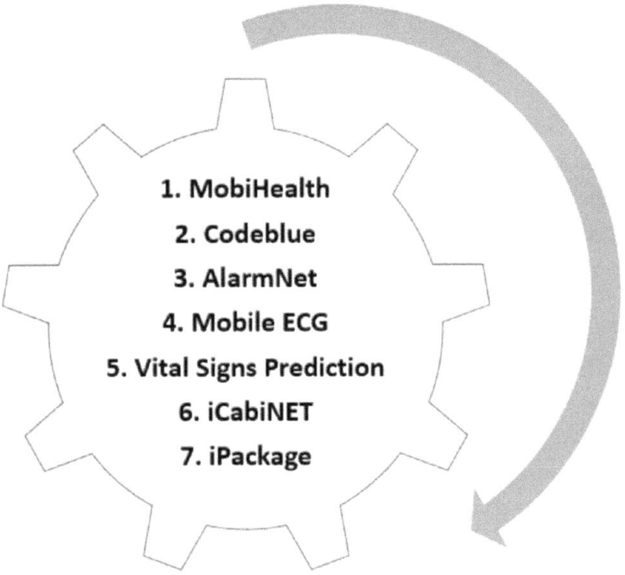

FIGURE 4.3 Devices used in MCPS.

4.3.1 MOBIHEALTH

The system is designed to collect data from smart wearables that monitor your medical status. In the event of an emergency, it also attempts to collect audio and video signals. The data is analyzed by smart sensors and sent to system servers. If there are any irregular readings, the servers are alerted. This can enhance response time and reaction during accidents or other emergency situations.

4.3.2 CODEBLUE

It is a tenet merged with the following biomedical sensors: (i) pulse oximeter oxygen saturation measurement device; (ii) two-lead ECG; and (iii) motion sensor with pub-sub-based routing framework-based software architecture. Its job is to efficiently manage and communicate among medical devices [19]. This system promotes the early use of smart routing and in-network aggregation.

4.3.3 ALARM-NET

It is a wireless system of bio-sensory devices that illustrates networks consisting of the following: (i) pulse rate; (ii) heart rate; (iii) oximetry evaluation; and (iv) ECG. Spatial contextual data is provided through environmental parameters such as temperature and humidity. The system also takes into consideration privacy, power management, and queries. It utilizes a pathway between the information collection and the storage unit [20]. Additionally, this system includes a GUI to assist healthcare professionals in determining the patient's vital signs.

4.3.4 MOBILE ECG

This ECG architecture utilizes smartphones as an interface and a data collector, while a home device serves as a measurement tool for ECG and analysis. The smartphones receive ECG data from mobile ECG recording devices via Bluetooth. If deemed necessary, the smartphones analyze and transmit the data to medical professionals [21]. The graphical interface also has the capability to display ECG readings and pulse rate.

4.3.5 VITAL SIGNS PREDICTION

This ECG architecture utilizes smartphones as an interface and a data collector, while a home device serves as a measurement tool for ECG and analysis. The smartphones receive ECG data from mobile ECG recording devices via Bluetooth. If deemed necessary, the smartphones analyze and transmit the data to medical professionals [21, 22]. The graphical interface also has the capability to display ECG readings and pulse rate.

4.3.6 MEDICINES CONSUMPTION APPLICATION

It can take care of people when they face analytical disabilities.

4.3.6.1 ICABINET

Smart RFID packaging is utilized in the case of iCabiNET, which detects the extraction of a tablet when an electrical flow is broken down in the RFID circuit. The system of medicine intake is observed using the RFID reader through a residential network available at homes [23]. If you use smart appliances with this system, you can also monitor the necessity to purchase medications that are running low in your supplies. Additionally, it can remind you to take your medications on time and with accurate dosage [24].

4.3.6.2 IPACKAGE

Different from RFID packaging, this is a prototype that features intelligent packaging, proposed by Pang et al. This system monitors medication intake along with vital signs. The system utilizes an array of controlled delamination material (CDM) films that incorporate unified control circuits. The CDM film consists of three layers: aluminum lower and upper layers, and a sticking mid-layer of electrochemical epoxy [25]. An electrochemical reaction occurs within the mid-layer when a voltage greater than a specific energy is applied to the lower (bottom) layer [26]. All of these technicalities ensure that the iPackage can only be accessed through an essential governing device, which also authorizes power over the dosage to prevent accidents or misuse.

4.4 CHALLENGES IN MCPS

We will discuss the challenges in CPS and then move on to the challenges faced in MCPS. Challenges in CPS range from the lack of CPS standards to the architecture of CPS [27]. Figure 4.4 provides an overview of the challenges faced in MCPS as well as CPS. We can gain a better understanding of the challenges in MCPS only when we have an idea of the challenges faced in CPS.

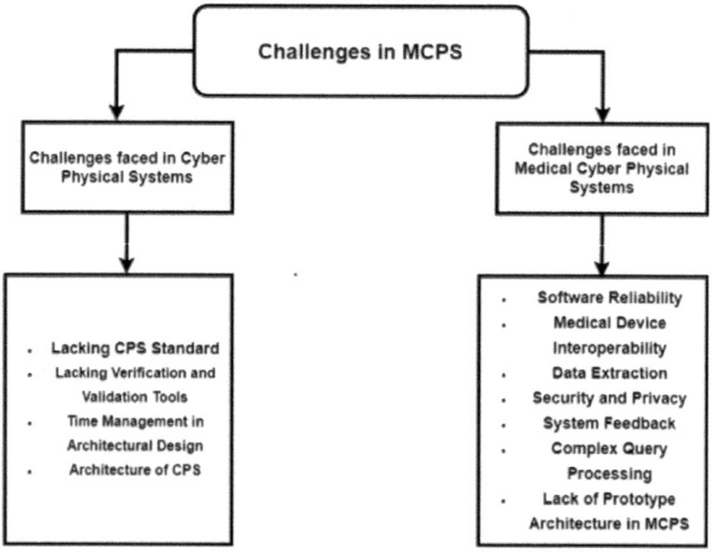

FIGURE 4.4 Challenges in MCPS.

4.4.1 CHALLENGES FACED IN CYBER PHYSICAL SYSTEMS (CPS)

4.4.1.1 LACKING CPS STANDARD

Identifying the information-based explanations and calculations for theories such as security, privacy, and safety is a scientifically challenging task [28]. It brings uncertainty in how the utilization and further reasoning will be carried out, even if the concepts are identified. For example, even professionals are entrusted with patient-doctor secrecy and the data privacy of patients [29]. A question arises as to what level of concern privacy will be considered breached. This clearly highlights how the CPS standard lacks in modern-day situations where privacy is of utmost importance.

4.4.1.2 LACKING CONFIRMATION AND TOOLS OF VALIDATION

We can utilize a CPS application to integrate the Wireless Sensor Network (WSN) and cloud computing architecture. Despite the availability of WSN and Cloud Computing modeling and simulation tools, there is a need for new simulation tools to comprehensively design the CPS system. These tools should be capable of handling the gathering, analysis, processing, and activation of heterogeneous sensors, communication layers, and other data types within the system [30]. This ensures the system remains integrated and responsive.

4.4.1.3 MANAGEMENT OF TIME IN ARCHITECTECTURAL DESIGN

In CPS, time management is a crucial issue due to the application of cloud integration and heterogeneous sensors. Given the need for real-time applications in MCPS, time management takes top priority. This includes time scaling, time synchronization, determining reference time, and interacting with various sensor devices while considering the characteristics of time [31]. An error in time management can have disastrous consequences, as lives may be at stake. Therefore, time management must be flawless in every single case.

4.4.1.4 ARCHITECTURE OF CPS

The challenges faced in CPS arise from the complexity of computing and physical dynamics. These challenges include time management, data correctness, and standards. The need of the hour is a common or consolidated architecture that can address all the aforementioned issues. Such an architecture will play a crucial role in the future of CPS design.

4.4.2 CHALLENGES FACED IN MCPS

4.4.2.1 RELIABILITY OF SOFTWARE

The program (software) is an important asset of healthcare devices, which ensures the functionality of devices and other functions. It also manages the proper coordination between patients and medical devices. Therefore, the safety and efficient management rely on proper software design and development. The system also improves when the software is more responsive, maintains a good history of records, and remains accurate at all times. With everything falling into place, a perfect healthcare system can be achieved.

4.4.2.2 INTEROPERABILITY OF MEDICAL DEVICES

Various medical devices are supposed to have distinct communication interfaces for smooth communication. We require a management system to oversee this and ensure that all the diverse medical devices are protected, reliable, and verified. Interoperability enables devices to function better and prevents accidents and emergency situations caused by haphazard communication. This managed approach will ensure effectiveness and avoid any adverse effects on the users, whether they are healthcare providers or patients.

4.4.2.3 EXTRACTION OF DATA

The collection of data from multiple patients by medical devices through various physiological criteria helps provide a wide variety of basic information about patients. It also helps in preventing future sickness and is effective in nullifying emergency situations. Although designing a device that seamlessly extracts physiological data from patients is challenging, it would be a revolutionary device. Many such devices are already in development.

4.4.2.4 SECURITY AND PRIVACY

Privacy is the foremost thing to be considered when it comes to organizing and handling patient data. The data of patients is blindly trusted with the healthcare system, so the system is expected to be trustworthy. Forfeiture of private data and details can even lead to mental unrest and abuse, which can further damage a patient's health. Privacy can be improved by implementing end-to-end encryption in the system. Encryption ensures that there is no third party who might tamper with or misuse the data. This can ensure proper functioning and smoothness in the system. Many systems currently in use rely on offline files and paperwork, which can be messy. Adopting organized ways can enhance security and patient satisfaction.

4.4.2.5 SYSTEM FEEDBACK

Developing a feedback system becomes challenging for MCPS due to the lack of design stability without the feedback system or improvement of patient care. MCPS has implemented a robust feedback system using a quick-witted alarm system. This alarm device is of utmost importance in alerting the care provider for potential threats, illnesses, or any emergency scenarios. However, the alarm system faces several challenges such as determining the types of physiological parameters, dealing with structural complexity, and ensuring effective implementation to enhance overall functionality. It is necessary to carefully select the relevant physiological variables for planning the alarm system, considering the wide range of illnesses. This task can be exhausting due to the need to detect specific sicknesses from various probabilities [32].

4.4.2.6 COMPLEX QUERY PROCESSING

Wireless sensors operated on batteries pose challenges due to the availability of heterogeneous biosensors, including issues such as lower power consumption and limited refining capabilities that hinder compound query processing. Complex Query Processing is very helpful in reducing the amount of transmission and enabling context-aware predictions, which in turn lowers power consumption in a limited energy arrangement. We can use this approach to access various physiological variables and pre-calculate possible sickness. However, handling this approach requires a complex design and algorithmic skill set.

4.4.2.7 LACK OF PROTOTYPE ARCHITECTURE IN MCPS

In the current scenario, the architecture lacks safe and dependable prototypes for experimentation, evaluation, and system enhancements that include devices in the health sector. This is yet another factor that contributes to the inability to ensure the correctness of the MCPS architecture in this uncertain environmental scenario.

4.5 ADVANTAGES OF MCPS

There are numerous advantages of technologies in the modern world, and a few of the advantages of MCPS [33] are listed in subsections. Figure 4.5

illustrates how technologies have affected and brought about a boom in the healthcare sector. Now, let's discuss the advantages of MCPS, which include network integration and various other benefits.

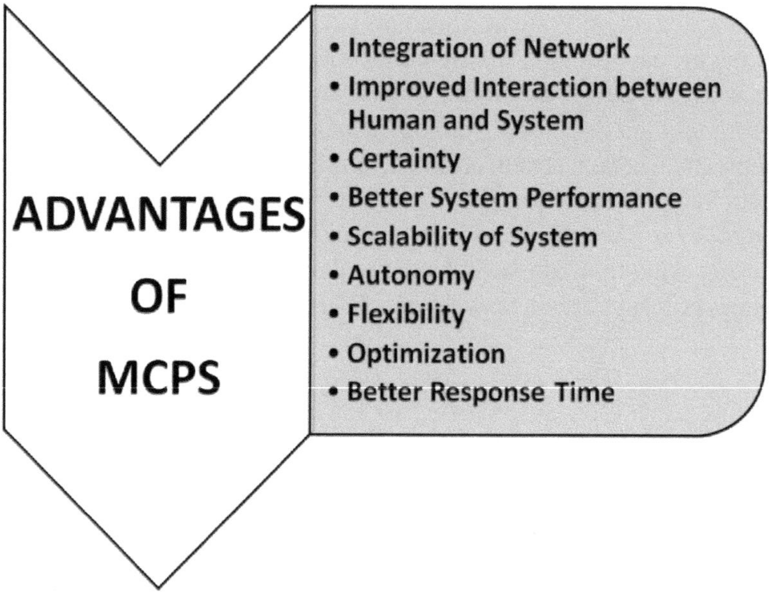

FIGURE 4.5 Advantages of MCPS.

4.5.1 INTEGRATION OF NETWORK

CPS is said to have integration with WSNs and cloud computing, which can offer consistency in associating levels. Various algorithmic platforms communicating over communication networks are also included in CPS. Integration of networks also improves the overall system communication.

4.5.2 IMPROVED INTERACTION BETWEEN HUMAN AND SYSTEM

It is necessary to comprehend the needs and design the structure accordingly. The adaptability to surrounding changes in variables that are critical for decision making is also enhanced. Thus, it also becomes essential for complex and dynamic systems. The interaction is made easier through this integrated system as people are often difficult to structure through independent systems.

4.5.3 CERTAINTY

Certainty involves the process that provides evidence that the design is justifiable and can be trusted. The proof can include past evidence or experiments in mimics and samples. CPS is structured to be capable of cooperating with new and unpredictable environments and functioning properly. It is expected to detect any unknown behavior by the system in order to study and improve the system.

4.5.4 BETTER SYSTEM PERFORMANCE

CPS provides enhanced system performance in terms of response and robotic redesign, thanks to the close interactivity between sensor devices and networked structure. Better assets and virtual subroutines ensure the availability of various sensing entities and various communication mechanisms, ultimately ensuring greater organizational execution by the CPS.

4.5.5 EXPANDABILITY OF SYSTEM

CPS has the ability to expand structures according to demands, utilizing the resources of Cloud Computing appropriately. People can access essential infrastructure without requiring additional resources, which presents a great opportunity. CPS is inherently heterogeneous as it integrates physical kinetics with computational procedures. This ensures that demands are met in a timely manner, with efficiency and accuracy.

4.5.6 AUTONOMY

CPS is capable of providing autonomy as it includes sensor-cloud integration, which is essentially a closed-loop system. This allows sensor devices to perform calculations of physical kinetics, which are then utilized in the cyber subsystems. These subsystems ultimately drive the devices and apps, which in turn affect physical procedures.

4.5.7 FLEXIBILITY

Few existing systems based on the CPS provide greater flexibility compared to previous research efforts in the fields of WSN and cloud computing. CPS systems offer stability, improved functionality, and faster response times. This, in turn, enhances the overall flexibility of the CPS system.

4.5.8 OPTIMIZATION

The biomedical sensors and cloud infrastructure that we currently have provide better optimizations for a variety of applications.

4.5.9 BETTER RESPONSE TIME

The faster communication and processing abilities of the sensors and cloud infrastructure have led to better reaction times, offering early detection of any remote failures and the appropriate implementation of shared resources like bandwidth. Better response times also ensure security, and areas of improvement can be easily considered. When you know the areas in which you lag, you can improve the system for the greater good.

4.6 CONCLUSION AND FUTURE SCOPE

This chapter presents a detailed analysis of CPS in the healthcare sector. To ensure proper comprehension of MCPS, a recapitulation of CPS in a generalized manner is offered first. A taxonomy of CPS and MCPS is also provided, including eight different components. The classification above proves resourceful in realizing the trends, methods, and possible solutions related to certain apps in CPS. This will not only improve the faulty areas in the pre-existing system but also provide a horizon that reflects the future possibilities in CPS and MCPS.

The future scope of CPS in healthcare includes:

1. **AI Healthcare:** Artificial intelligence (AI) is a booming trend that can be used efficiently when merged with the CPS system in

healthcare. AI's accuracy and intelligence can effectively manage emergency situations and even suggest solutions for them.

2. **Mobile Healthcare Services:** Vehicles installed with advanced MCPS systems can have a significant impact on the development of the healthcare sector. People who suffer or are affected by the lack of services in remote areas can access this system through these mobile healthcare services.
3. **Smart Homes:** It is a really good way of taking care of yourself or your loved ones. They will experience assisted living, and the instructions fed to the system will allow it to manage their health and provide timely reminders for medication consumption, routine checkups, and even call an ambulance in case of emergencies.

KEYWORDS

- **cyber physical system**
- **electronic medical records**
- **Food and Drug Administration**
- **iCabiNET**
- **medical cyber physical systems**
- **medical science**
- **social network**

REFERENCES

1. Sonntag. (n.d.). Medical cyber-physical systems. *Cyber-Physical System Design with Sensor Networking Technologies*, 311–333. https://doi.org/10.1049/pbce096e_ch13.
2. Finkbeiner, B., Keller, A., Schmidt, J., & Schwenger, M., (2021). Robust monitoring for medical cyber-physical systems. *Proceedings of the Workshop on Medical Cyber-Physical Systems and Internet of Medical Things*. https://doi.org/10.1145/3446913.3460318.
3. Li, T., Cao, J., Liang, J., & Zheng, J., (2014). Towards context-aware medical cyber-physical systems: Design methodology and a case study. *Cyber-Physical Systems*, *1*(1), 5–23. https://doi.org/10.1080/23335777.2014.972686.

4. Ivanov, R., Weimer, J., & Lee, I., (2018). Context-aware detection in medical cyber-physical systems. In: *2018 ACM/IEEE 9th International Conference on Cyber-Physical Systems (ICCPS)*. https://doi.org/10.1109/iccps.2018.00030.
5. Tiwari, A., Sharma, N., Kaushik, I., & Tiwari, R., (2019). Privacy issues & security techniques in big data. In: *2019 International Conference on Computing, Communication, and Intelligent Systems (ICCCIS)*. doi: 10.1109/icccis48478.2019.8974511.
6. Tyagi, N., Gautam, S., Goel, A., & Mann, P., (2021). A framework for blockchain technology including features. In: Hassanien, A. E., Bhattacharyya, S., Chakrabati, S., Bhattacharya, A., & Dutta, S., (eds.), *Emerging Technologies in Data Mining and Information Security: Advances in Intelligent Systems and Computing* (Vol. 1286). Springer, Singapore. https://doi.org/10.1007/978-981-15-9927-9_62.
7. Sharma, N., Kaushik, I., Bhushan, B., Gautam, S., & Khamparia, A., (2020). Applicability of WSN and biometric models in the field of healthcare. *Deep Learning Strategies for Security Enhancement in Wireless Sensor Networks Advances in Information Security, Privacy, and Ethics*, 304–329. doi: 10.4018/978-1-7998-5068-7.ch016.
8. Nitin, T., Bharat, B., Siddharth, G., Nikhil, S., & Santosh, K., (2021). 9 an integrated approach of blockchain & big data in health care sector. In: *Applications of Machine Learning in Big-Data Analytics and Cloud Computing* (pp.183–205). River Publishers.
9. Kathuria, R. S., Gautam, S., Singh, A., Khatri, S., & Yadav, N., (2019). Real-time sentiment analysis on Twitter data using deep learning (Keras). In: *2019 International Conference on Computing, Communication, and Intelligent Systems (ICCCIS)* (pp. 69–73). doi: 10.1109/ICCCIS48478.2019.8974557.
10. Malik, A., Gautam, S., Khatoon, N., Sharma, N., Kaushik, I., & Kumar, S., (2020). Analysis of black-hole attack with its mitigation techniques in ad-hoc network. *Deep Learning Strategies for Security Enhancement in Wireless Sensor Networks Advances in Information Security, Privacy, and Ethics,* 211–232. doi: 10.4018/978-1-7998-5068-7.ch011.
11. Harjani, M., Grover, M., Sharma, N., & Kaushik, I., (2019). Analysis of various machine learning algorithms for cardiac pulse prediction. In: *2019 International Conference on Computing, Communication, and Intelligent Systems (ICCCIS*). doi: 10.1109/icccis48478.2019.8974519.
12. Kathuria, R. S., Gautam, S., Singh, A., Singh, A., & Yadav, N., (2021). Sentiment analysis on images using convolutional neural network. In: Hassanien, A. E., Bhattacharyya, S., Chakrabati, S., Bhattacharya, A., & Dutta, S. (eds.), *Emerging Technologies in Data Mining and Information Security; Advances in Intelligent Systems and Computing* (Vol. 1300). Springer, Singapore. https://doi.org/10.1007/978-981-33-4367-2_20.
13. Plourde, J., Arney, D., & Goldman, J. M., (2014). OpenICE: An open, interoperable platform for medical cyber-physical systems. In: *2014 ACM/IEEE International Conference on Cyber-Physical Systems (ICCPS)*. https://doi.org/10.1109/iccps.2014.6843734.

14. Mahmoud, M. S., & Hamdan, M. M., (2019). Improved control of cyber-physical systems subject to cyber and physical attacks. *Cyber-Physical Systems, 5*(3), 173–190. https://doi.org/10.1080/23335777.2019.1631889.
15. Konstantinou, Charalambos, Michail Maniatakos, Fareena Saqib, Shiyan Hu, Jim Plusquellic, & Yier Jin (2015). Physical infrastructures for cyber-physical systems. *Cyber-Physical Systems* (pp. 1, 2). https://doi.org/10.1201/b19206-2.
16. Sharma, A., Sharma, N., Kaushik, I., Kumar, S., & Khatoon, N., (2020). Predictive analysis of type 2 diabetes using hybrid ml model and IoT. *IoT Security Paradigms and Applications*, 303–320. doi: 10.1201/9781003054115-14.
17. Goyal, S., Sharma, N., Bhushan, B., Shankar, A., & Sagayam, M., (2021). IoT enabled technology in secured healthcare: Applications, challenges, and future directions. In: Hassanien, A. E., Khamparia, A., Gupta, D., Shankar, K., & Slowik, A., (eds.), *Cognitive Internet of Medical Things for Smart Healthcare, Studies in Systems, Decision and Control,* (Vol. 311). Springer, Cham. https://doi.org/10.1007/978-3-030-55833-8_2.
18. Alamelu, J. V., Priscilla Dinkar Moyya, & Mythili Asaithambi. "Medical Cyber Physical System Architecture for Smart Medical Pumps." In *Deep Learning Applications for Cyber-Physical Systems*, pp. 207–221. IGI Global, 2022.
19. Sharma, N., Kaushik, I., Rathi, R., & Kumar, S., (2020). Evaluation of accidental death records using hybrid genetic algorithm. *SSRN Electronic Journal*. doi: 10.2139/ssrn.3563084.
20. Stojmenovic, I., & Zhang, F., (2014). Inaugural issue of 'cyber-physical systems.' *Cyber-Physical Systems, 1*(1), 1–4. https://doi.org/10.1080/23335777.2015.970764.
21. Goel, A., Bhushan, B., Tyagi, B., Garg, H., & Gautam, S., (2021). Blockchain and machine learning: Background, integration challenges and application areas. In: Tavares, J. M. R. S., Chakrabarti, S., Bhattacharya, A., & Ghatak, S., (eds.), *Emerging Technologies in Data Mining and Information Security; Lecture Notes in Networks and Systems* (Vol. 164). Springer, Singapore. https://doi.org/10.1007/978-981-15-9774-9_29.
22. Mann, P., Tyagi, N., Gautam, S., & Rana, A., (2020). Classification of various types of attacks in IoT environment. In: *2020 12th International Conference on Computational Intelligence and Communication Networks (CICN)* (pp. 346–350). doi: 10.1109/CICN49253.2020.9242592.
23. King, A. L., Feng, L., Sokolsky, O., & Lee, I., (2013). Assuring the safety of on-demand medical cyber-physical systems. In: *2013 IEEE 1st International Conference on Cyber-Physical Systems, Networks, and Applications (CPSNA)*. https://doi.org/10.1109/cpsna.2013.6614238.
24. Rathi, R., & Kumar, S. (2013). Mobile Cyber-Physical Systems. *Cyber-Physical Systems* (pp. 55–68). https://doi.org/10.1201/b15552-7.
25. Gautam, S., Malik, A., Singh, N., & Kumar, S., (2019). Recent advances and countermeasures against various attacks in IoT environment, In: *2019 2nd International Conference on Signal Processing and Communication (ICSPC)* (pp. 315–319).
26. Yu, Wenjin, Tharam Dillon, Fahed Mostafa, Wenny Rahayu, & Yuehua Liu (2018). Bringing cyber-physical systems into industrial practice,*2018 IEEE Industrial Cyber-Physical Systems (ICPS)*. https://doi.org/10.1109/icphys.2018.8390758.

27. Goel, A., Gautam, S., Tyagi, N., Sharma, N., & Sagayam, M., (2021). Securing biometric framework with cryptanalysis. *Intelligent Data Analytics for Terror Threat Prediction,* 181–208. https://doi.org/10.1002/9781119711629.ch9.
28. Srinivasa, Krishnarajanagar GopalaIyengar, Nabeel Siddiqui, Abhishek Kumar, & Anita Kanavalli. (2015). Cyber-physical systems: Beyond sensor networks and embedded systems. *Cyber-Physical Systems,* 98–123. https://doi.org/10.1201/b19206-8.
29. Rustagi, A., Manchanda, C., Sharma, N., & Kaushik, I., (2020). Depression anatomy using combinational deep neural network. *Advances in Intelligent Systems and Computing International Conference on Innovative Computing and Communications,* 19–33. doi: 10.1007/978-981-15-5148-2_3.
30. Bordel, Borja, Ramón Alcarria, Tomás Robles, & Diego Martín (2015). Wireless sensor networks toward Cyber-physical systems. *Cyber-Physical Systems,* (pp. 124–145). https://doi.org/10.1201/b19206-9.
31. Rathi, R., Sharma, N., Manchanda, C., Bhushan, B., & Grover, M., (2020). Security challenges & controls in cyber-physical system. In: *2020 IEEE 9th International Conference on Communication Systems and Network Technologies (CSNT).* doi: 10.1109/csnt48778.2020.9115778.
32. Rana, A., Gautam, S., Mann, P., & Tyagi, N., (2020). Classification for big data tools and its challenges and issues. *SSRN Electronic Journal.* https://doi.org/10.2139/ssrn.3746970.
33. Tanik, U. J., & Begley, A., (2013). An adaptive cyber-physical system framework for cyber-physical systems design automation. *Applied Cyber-Physical Systems,* 125–140. https://doi.org/10.1007/978-1-4614-7336-7_11.

PART II

Intelligent Decision-Making and Analysis Technology

CHAPTER 5

SIGNIFICANCE OF ML AND NLP IN SENTIMENT ANALYSIS

PUMMY DHIMAN,[1] AMANDEEP KAUR,[1] ANUPAM BONKRA,[2] and NIDHI CHAHAL[2]

[1]*Chitkara University, Punjab, India*

[2]*Chandigarh Engineering College, Punjab, India*

ABSTRACT

With the introduction of Web 2.0, sentiment analysis (SA) has expanded dramatically. The internet has provided a platform for people to express their opinions. Nowadays, individuals freely share their opinions on various topics online, and these reviews play a significant role in decision-making processes. The objective of SA is to utilize this data to gather crucial information about popular opinions, which can be used to enhance company choices, political campaigns, and consumer behavior. Consequently, SA has emerged as a new research area. In this study, a publicly available Twitter dataset was used and processed using NLTK to demonstrate the importance of the AI approach and machine learning (ML) in SA. Three classifiers were employed to binary classify the given input. Experimental results indicate that SVC achieves the highest accuracy.

5.1 INTRODUCTION

Sentiment refers to a person's expression, emotions, attitude, or any opinion [1]. Nowadays, almost everyone exploits the internet or social

The Fusion of Artificial Intelligence and Soft Computing Techniques for Cybersecurity.
M. A. Jabbar, Sanju Tiwari, Subhendu Kumar Pani, & Stephen Huang (Eds.)

media platforms to articulate their opinions on various matters, such as products, software, applications, movies, web series, and more. These opinions have become increasingly important as social networking sites, business models, and product managers utilize feedback forms to collect valuable customer feedback and analyze it for making better business decisions. Sentiment analysis (SA) is the process of understanding the text and identifying the user's perspective. Through SA, text can be categorized into different sentiments, whether they are positive, negative, or neutral. This type of text classification is used by data analysts, business analysts, and others to quickly analyze collected data and feedback. There are three main methods for sentiment categorization: lexicon-based, machine learning (ML)/deep learning (DL), and hybrid [2], as shown in Figure 5.1. SA is a technique used to analyze a script or text in order to extract thoughts, opinions, and subjectivity, and determine the polarity [3], whether it is positive, negative, or neutral [1]. Opinion mining is another term for SA [4].

5.1.1 LEXICON-BASED APPROACH

Three approaches are available for SA: the supervised method outperforms the lexicon-based method in terms of performance, and the hybrid approach combines the two. In the Lexicon approach, extracted expressions from the text are compared to its dictionary to provide a label. This method explains the (positive, negative, and neutral) polarity of the text content using a sentiment lexicon [5]. The attitudes of Twitter can be detected using various NLP techniques, including Stop Word Removing, Parts of Speech Tagging, name entity recognition (NER), and a large number of words, etc. Dictionaries are used as references in these methods. No training is required, hence less computing power is needed. With the aid of dictionaries, we are employing the lexicon approach to divide the text into two categories: "positive" and "negative" [24]. A dictionary-like collection of words, or a lexicon, is one where the words are grouped alphabetically. A dictionary-based method and a corpus-based approach are two variations of this strategy. Here, we are employing a corpus-based methodology. A corpus is a significant collection of words or text that analyzes the relationships between different languages and formulates a set of conceptual rules that govern a natural language from texts in that language [25].

5.1.2 ML/DL APPROACH

Deep learning (DL) and semantic ML both aim to create AI that mimics human capabilities such as indexing, verifying, predicting, and reasoning. Both disciplines have had a significant impact on the analysis of data and knowledge, as well as the related conceptual formulations. ML/DL uses a train-test split mechanism and pre-defined algorithms to classify sentiments.

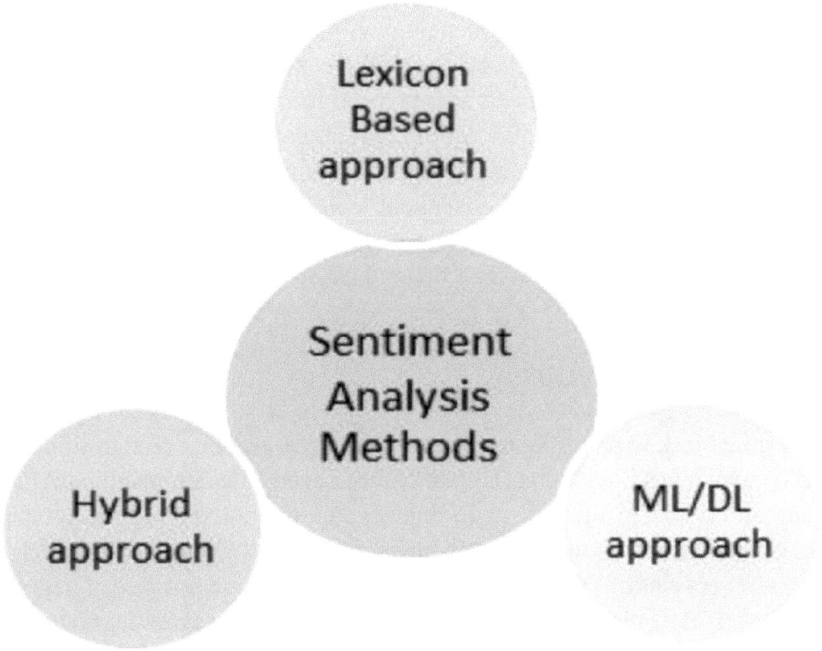

FIGURE 5.1 Sentiment categorization methods.

5.1.3 HYBRID APPROACH

The hybrid method of SA employs both statistical and knowledge-based techniques for polarity detection. It benefits from the stability of the lexicon-based strategy and the superior accuracy of ML (statistical approaches). The process begins with a hybrid standard classification, which is then refined into a sophisticated technique that includes linguistic cataloging of semantic divergence modeled through fuzzy sets. The holistic approach to SA utilizes both statistical and knowledge-based methods for polarity identification. It benefits from the stability of the lexicon-based strategy and the superior accuracy of ML [6, 7].

5.2 MOTIVATION

As we know, consumers and producers are mutually dependent. Consumer feedback is crucial in making better business decisions for producers [8]. However, analyzing people's opinions is not an easy task. Sentiments also play a major role in influencing people's minds. The availability of social platforms is being used to organize mass gatherings and oppose oppression. Social media can also be used to incite hatred against a class of people or a specific ethnic group, which can lead to a significant loss of life. Therefore, it is essential to have SA systems that can detect and mitigate such phenomena. Consequently, SA systems capable of detecting such phenomena are required. The goal of this study is to use ML techniques to predict the labels on a given dataset.

5.3 PREVIOUS WORK

A critical task in natural language processing is emotion recognition from text, which has huge potential to advance several fields, including AI and human-computer interaction. In this chapter, the authors suggest labeling a large number of tweets according to their underlying sentiment. The researchers classify the emotions of an expression into positive or negative using DL algorithms. They conducted experiments on three separate datasets, employing recurrent neural networks (RNNs) and long short-term memory to demonstrate high generalization ability and achieved an accuracy of 91.3% for both subclasses [9]. This chapter depicts the advancements in SA in Indian languages. The research compares SA carried out in various Indian languages, along with categorization methods based on lexicon, dictionary, and ML. Additionally, it lists the lexical resources that can be used for SA of Indian languages and discusses the challenges in creating such resources with limited availability [10]. This study contributes to the field of SA for the classification of consumer reviews, which is beneficial for analyzing a large amount of tweets with highly complex opinions that are either positive, negative, or neutral [11]. The goal of this effort is to develop a classifier using ML techniques that can predict the polarity of a comment. Their study consisted of three main steps: data extraction, processing, and modeling. They utilized the NLTK dataset to construct their model. The variables were then generated and processed using text mining algorithms. They conducted two experiments to assess the effectiveness of their model after creating a classifier to categorize tweets based on affirmative and pessimistic feelings [12]. This latest research focused on analyzing tweets related to the Uri massacre to evaluate sentiments and polarity. Manuscript removal algorithms were used to extract tweets for passion and polarization. A collection of frequently used words was created by preprocessing and recoding 5,000 tweets. R was implemented in this study to analyze polarity and emotions, with 94.3% of the research participants expressing loathing towards the Uri attack [13]. The goal of this chapter is to examine Indian citizens' opinions on the government's nationwide lockdown, which was implemented to curb the spread of the coronavirus. SA of tweets made by Indian residents was carried out using NLP and ML classifiers. A total of 12,741 tweets containing the phrase "India lockdown" were retrieved between April 5, 2020, and April 17, 2020. The Linear SVC classifier and unigrams were used in the experiment, achieving a precision of 84.4%. According to the study's findings, the majority of the Indian population agrees with the government's decision to impose a lockdown during the coronavirus outbreak [2]. Analyzing sentiment in raw text becomes particularly challenging when dealing with sarcastic text or the presence of both optimistic and pessimistic opinions in the same text. This model can be applied to various business models and provides an accuracy and precision of about 95% [1]. The study's first finding is that an increase in tweet volume will inevitably lead to longer execution time for the algorithm. The second finding is that RF takes significantly longer to execute compared to SVM and NB. Therefore, this literature study systematically investigates and evaluates corporate and community white papers, academic research articles, journals, and reports to examine and analyze the concept of SA [14]. This study contributes to the field of SA for the classification of consumer reviews, which is beneficial for analyzing data from a large number of tweets with highly unstructured opinions that are either positive, negative, or neutral [12]. A technique called latent semantic analysis (LSA) examines text using precise mathematical computations and analyzes the connections between terms in the documents and between the documents in the corpus. LSA demonstrates that a single word can have multiple meanings, compress several concepts with the same semantics, and represent texts in a conceptual space with fewer dimensions [15]. Natural language processing (NLP) is the technique used in Twitter SA. In this project called Twitter SA, we explore the sentiments expressed in each tweet. We recognize three types of sentiments: positive, neutral, and negative. Initially, analyzing the emotions underlying each tweet posed the biggest challenge, but ML can now address this issue [21]. SA can be used to evaluate the positivity, negativity, or neutrality of written communication. People use various social media channels, such as Facebook and Twitter, where opinions can be accurately assessed.

This study utilizes a variety of ML techniques. Twitter is employed to analyze untrained data, which is then compared to trained software capable of determining neutrality, negativity, or positivity. Effective SA can be conducted on social networking data by removing subjective information from written material [22]. Classifying image sentiments based on raw photos and elusive human emotions can be a challenging task for computers. There is ample room for research in the field of image sentiment classification, especially in social data analytics. Nowadays, individuals often express their thoughts on the internet by uploading photographs to various social media platforms such as Facebook, Instagram, Twitter, and others. This research employs DL approaches to explore the latest advancements in image SA. In addition to DL algorithms, traditional ML approaches are also compared. The study suggests that a combination of rapid recurrent neural network and CNN can achieve high accuracy with minimal temporal complexity. According to previous studies [23], CNN achieves an average accuracy of approximately 96.50% in classifying emotions in a flicker picture collection.

5.4 METHODOLOGY

The following tools and methodology are utilized to accomplish the objective of SA. This research work used Python as the programming language, Anaconda as the programming environment, and various associated libraries such as Pandas, NumPy, NLTK, Scikit-learn, Matplotlib, etc. Figure 5.2 represents the research methodology flow. The research model will describe the different sections: (i) dataset collection; (ii) preprocessing; (iii) ML classifier; and (iv) performance metric evaluation.

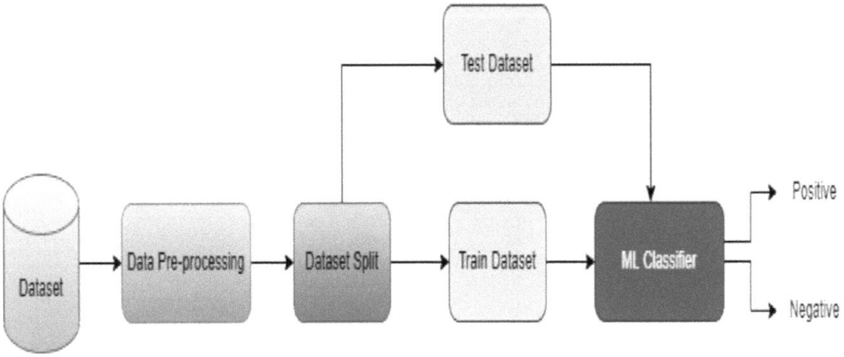

FIGURE 5.2 Methodology.

5.4.1 DATASET

The dataset utilized in this study consists of 31,962 labeled tweets in CSV format. It is available in the public domain and includes entries for tweet ID, label, and the tweet itself. This dataset contains short-length tweets, with a label of '1' indicating a racist (positive) tweet and '0' indicating a non-racist (negative) tweet.

5.4.2 SOFTWARE-HARDWARE REQUIREMENT

Intel(R) Core(TM) i3-5005U CPU @ 2.00 GHz 2.00 GHz, 8 GB RAM, Windows OS, Anaconda IDE, and Python version 3.10 were used to carry out this study. The experiment utilized Pandas, NumPy, Matplotlib, NLTK, and re libraries.

5.4.3 PRE-PROCESSING

Preprocessing raw sentiment texts was necessary before feeding them into the models. The first step was to eliminate stop words. Afterwards, words are spelled correctly to clean up the dataset. Throughout the text, white space is used to separate each paragraph, and words are stemmed to remove suffixes. Finally, the word tokens are separated with white space to present a clean corpus, which had been tokenized later for use in the models. As shown in Figure 5.3, it is preprocessed using the NLTK Python library to deal with NLP [16] tasks. Certain patterns are removed in the next step using the sub() method, and then the removal of Twitter handles is managed.

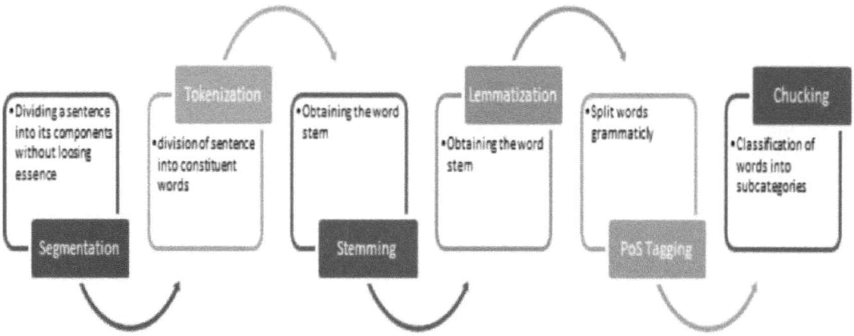

FIGURE 5.3 NLP process.

After this, if there are any special characters present in the given strings, they have been removed using NLTK methods [17]. Removal of shorter words, less than 3 characters, has taken place. Tokenization, i.e., dividing the given string into tokens, is taken care of in the next step. Later on, the stem() method is used for stemming purposes. To find out various patterns, exploratory data analysis (EDA) is also done. It is necessary to vectorize the data after the data cleaning step has been completed. The text is transformed into a numerical two-dimensional matrix through vectorization in order to reduce time complexity [18].

After the pre-processing step, the dataset needs to be split into two parts, namely the train-test split. ML algorithms are trained on datasets (called training data) and optimized to find specific patterns or outputs in the data. Rules and data structures are the components of a learned ML model. The first part is for training purposes, i.e., to build the model, and the next split is for testing the built model. In this case, 75% of the data is used for training and 25% for testing. The experiment uses tweets and labels. After pre-processing, the data is fed into supervised ML classifiers [19, 20]. A ML algorithm is a mathematical method for detecting patterns in data sets. Statistics, calculus, and linear algebra are frequently used in ML algorithms. Linear regression, decision trees (DT), random forest (RF), SVM, and XGBoost are some popular ML algorithms. For this purpose, three supervised ML classifiers are used: logistic regression (LR), SVM, and decision tree:

i. **Logistic Regression (LR):** It is used to determine whether or not an input belongs to a specific group.

ii. **Support Vector Machines (SVMs):** These group objects according to their common features by generating coordinates in an n-dimensional space.

iii. **Decision Trees (DT):** These like classifiers, is used to determine which category an input belongs to by traversing the tree's leaves and nodes.

5.5 PERFORMANCE METRICS (PM)

The efficiency of the model was evaluated using the following performance parameters: accuracy, precision, recall, and F1-score. Based on Table 5.1, all these parameters are derived from the confusion matrix shown in Figure 5.4: true positives (TP), false positives (FP), true negatives (TN), and false negatives (FNs).

	Predicted	**Predicted**
Actual	**TP**	**FN**
Actual	**FP**	**TN**

FIGURE 5.4 Confusion matrix.

Table 5.1 gives the explanation of confusion matrix parameters.

TABLE 5.1 Parameter Explanation

Parameter	Explanation
TP	Positive prediction and actually it is also positive.
FP	Positive prediction but in actual it is negative.
TN	Negative prediction and actually it is also negative.
FN	Negative prediction but in actual it is positive.

i. **Accuracy (TN + TP/TP + FP + FN + TN)** = Identifies the positive predictions to the total number of predictions.

ii. **Precision (TP/TP + FP)** = Determines how often a modal predicts the positive class correctly.

iii. **Recall (TP/TP + FN)** = Identifies out of all positive labels, how many did the modal correctly identify.

iv. **F1-Score (2*(Precision*Recall)/(Precision + Recall)) =** Harmonic mean of precision and recall.

5.6 RESULT

This section addresses the following research question – how accurate are traditional ML models in analyzing sentiments in the given dataset? After pre-processing the dataset, ML algorithms were applied, and a supervised learning approach yielded satisfactory results for the classification of Twitter reviews. Table 5.2 reports the performances of various traditional ML models. For LR, the achieved TP is 7362, FP is 348, FN is 65, and TN is 215. For Decision Tree Classifier, TP is 7131, FP is 255, FN is 296, and TN is 308. For SVC, TP is 7394, FP is 372, FN is 33, and TN is 191.

TABLE 5.2 Result Analysis

	LR	DT	SVC
Accuracy	0.948310	0.931039	0.949312
Precision	0.767857	0.509934	0.852679
Recall	0.381883	0.547069	0.339254
F1-Score	0.510083	0.527849	0.485388

It is concluded here that SVC is better than other classifiers in achieving accuracy and precision. However, there is a need for improvement in recall and F1-score. The efficiency of the three models is graphically shown in Figure 5.5.

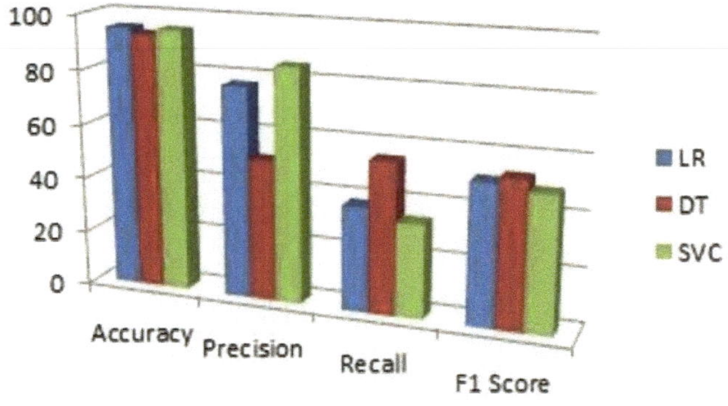

FIGURE 5.5 Graphical representation of results.

The efficiency of the three models is graphically shown in Figure 5.5. It is concluded that SVC is better than other classifiers in achieving accuracy and precision. However, there is a need for improvement in recall and F1-score.

5.7 CONCLUSION

In today's digital era, we have information at our fingertips. We engage in online shopping, watch web series, and share our opinions online to inform others about products. We also rely on reviews before making purchases or decisions. As a result, online reviews have sparked extensive research on sentiment analysis (SA) in recent years. These comments are written in human language, which computers struggle to understand. This is where NLP plays a significant role in bridging this gap. The Twitter SA dataset has been explored using ML models to classify tweets as positive or negative sentiment and compared in terms of accuracy. However, there is still room for improvement in achieving more efficient results by utilizing transformer-based techniques and hybrid approaches.

KEYWORDS

- **decision tree**
- **F1-score**
- **hybrid approach**
- **logistic regression**
- **machine learning**
- **natural language processing**
- **opinion**
- **sentiment analysis**
- **support vector classifier**

REFERENCES

1. Pati, S. P., & Pradhan, B. B., (2020). Comparison between machine learning algorithms used for sentiment analysis. *Int. J. Adv. Res. Eng. Technol., 11*(12), 220–228. doi: 10.34218/IJARET.11.12.2020.026.
2. Gupta, P., Kumar, S., Suman, R. R., & Kumar, V., (2021). Sentiment analysis of lockdown in India during COVID-19: A case study on Twitter. *IEEE Trans. Comput. Soc. Syst., 8*(4), 939–949. doi: 10.1109/TCSS.2020.3042446.
3. Wu, X., Cheng, S., & Chai, Z., (2017). *Fake News Stance Detection. 9*(1), 611–625.
4. Alonso, M. A., Vilares, D., Gómez-Rodríguez, C., & Vilares, J., (2021). Sentiment analysis for fake news detection. *Electron., 10*(11). doi: 10.3390/electronics10111348.
5. Sadia, A., Khan, F., & Bashir, F., (2018). An overview of lexicon-based approach for sentiment analysis. *Int. Electr. Eng. Conf., No. IEEC* (pp. 1–6).
6. Appel, O., Chiclana, F., & Carter, J., (2016). A hybrid approach to sentiment analysis. In: *2016 IEEE Congr. Evol. Comput., No. CCI* (pp. 4950–4957). doi: 10.1109/CEC.2016.7744425.
7. Gupta, I., & Joshi, N., (2020). Enhanced Twitter sentiment analysis using hybrid approach and by accounting local contextual semantic. *J. Intell. Syst., 29*(1), 1611–1625. doi: 10.1515/jisys-2019-0106.
8. Mohammad, S. M., (2021). Sentiment analysis. *Emot. Meas.,* 323–379. doi: 10.1016/b978-0-12-821124-3.00011-9.
9. Shilpa, P. C., Shereen, R., Jacob, S., & Vinod, P., (2021). Sentiment analysis using deep learning. *Proc. 3rd Int. Conf. Intell. Commun. Technol. Virtual Mob. Networks, ICICV 2021* (pp. 930–937). doi: 10.1109/ICICV50876.2021.9388382.
10. Shelke, M., & Deshmukh, S., (2020). Recent advances in sentiment analysis of Indian languages. *Int. J. Futur. Gener. Commun. Netw., 13*(4), 1656–1675.
11. Gautam, G., & Yadav, D., (2014). Sentiment analysis of Twitter data using machine learning approaches and semantic analysis. In: *2014 7th Int. Conf. Contemp. Comput. IC3 2014* (Vol. 12, No. 6, pp. 437–442). doi: 10.1109/IC3.2014.6897213.
12. Kalangi, R. R., Maloji, S., Tejasri, N., Chand, P. P., & Ch, V. P., (2021). Sentiment analysis using machine learning. *Proc. – 2021 3rd Int. Conf. Adv. Comput. Commun. Control Networking, ICAC3N 2021* (pp. 116–121). doi: 10.1109/ICAC3N53548.2021.9725499.
13. Kawade, D. R., & Oza, D., K. S., (2017). Sentiment analysis: Machine learning approach. *Int. J. Eng. Technol., 9*(3), 2183–2186. doi: 10.21817/ijet/2017/v9i3/1709030151.
14. Pandey, S., Tekchandani, H., & Verma, S., (2020). A literature review on application of machine learning techniques in pancreas segmentation. In: *2020 1st Int. Conf. Power, Control Comput. Technol. ICPC2T 2020* (pp. 401–405). doi: 10.1109/ICPC2T48082.2020.9071443.
15. Kherwa, P., & Bansal, P., (2018). Latent semantic analysis: An approach to understand semantic of text. *Int. Conf. Curr. Trends Comput. Electr. Electron. Commun. CTCEEC 2017*, 870–874. doi: 10.1109/CTCEEC.2017.8455018.

16. Rajput, A., (2019). Natural language processing, sentiment analysis, and clinical analytics. *Innov. Heal. Informatics A Smart Healthc. Prim.* (pp. 79–97). doi: 10.1016/B978-0-12-819043-2.00003-4.
17. Khurana, D., Koli, A., Khatter, K., & Singh, S., (2017). *Natural Language Processing: State of the Art, Current Trends and Challenges*. doi: 10.1007/s11042-022-13428-4.
18. Raza, H., Faizan, M., Hamza, A., Mushtaq, A., & Akhtar, N., (2019). Scientific text sentiment analysis using machine learning techniques. *Int. J. Adv. Comput. Sci. Appl., 10*(12), 157–165. doi: 10.14569/ijacsa.2019.0101222.
19. Wankhade, M., Chandra, A., Rao, S., & Kulkarni, C. (2021). *A Survey on Sentiment Analysis Methods, Applications, and Challenges,* 123AD. doi: 10.1007/s10462-022-10144-1.
20. Drus, Z., & Khalid, H., (2019). Sentiment analysis in social media and its application: Systematic literature review. *Procedia Comput. Sci., 161*, 707–714. doi: 10.1016/j.procs.2019.11.174.
21. Singh, S., (2020). Twitter sentiments analysis using machine learning. *Int. J. Sci. Res. Comput. Sci. Eng. Inf. Technol., 3307*, 312–320. doi: 10.32628/cseit206456.
22. Kalangi, R. R., Maloji, S., Tejasri, N., Chand, P. P., & Ch, V. P., (2021). Sentiment analysis using machine learning. In: *Proc. – 2021 3rd Int. Conf. Adv. Comput. Commun. Control Networking, ICAC3N 2021* (pp. 116–121). doi: 10.1109/ICAC3N53548.2021.9725499.
23. Salunke, V., & Panicker, S. S., (2021). Image sentiment analysis using deep learning. In: Ranganathan, G., Chen, J., & Rocha, Á., (eds.), *Inventive Communication and Computational Technologies. Lecture Notes in Networks and Systems* (Vol. 145). Springer, Singapore. https://doi.org/10.1007/978-981-15-7345-3_12.
24. Nigam, N., & Yadav, D., (2019). *Lexicon-Based Approach to Sentiment Analysis of Tweets Using R Language: Second International Conference, ICACDS 2018* (pp. 372–380). Dehradun, India. doi: 10.1007/978-981-13-1810-8.
25. Singh, V., Singh, G., Rastogi, P., & Deswal, D., (2018). Sentiment analysis using a lexicon-based approach. *PDGC 2018 – 2018 5th Int. Conf. Parallel, Distrib. Grid Comput.* (pp. 13–18). doi: 10.1109/PDGC.2018.8745971.

CHAPTER 6

INTELLIGENT INNOVATIVE SYSTEMS: EXPERIENCE IN MULTIDISCIPLINARY OUTLOOK ON CURRENT CHALLENGES AND CONTINGENCY

N. INDUMATHI,[1] M. SHANMUGA ESWARI,[2] R. REVATHY,[2] and R. RAMALAKSHMI[3]

[1]*Assistant Professor, Department of Computer Science and Engineering, Rajalakshmi Institute of Technology, Chennai*

[2]*Department of Computer Applications, Kalasalingam Academy of Research and Education, Krishnankoil, Tamil Nadu, India*

[3]*Department of Computer Science and Engineering, Kalasalingam Academy of Research and Education, Krishnankoil, Tamil Nadu, India*

ABSTRACT

Advanced technologies have practically reformed every aspect of human existence, including the implementation of machine learning (ML) innovations such as intelligent decision-making systems across various domains. In the present era, ML applications are being sought after for deployment in fields like medicine, industrial safety, and agriculture, with the aim of predicting future events. In the medical field, computer-aided decision support systems and machine intelligence can be easily utilized for disease

The Fusion of Artificial Intelligence and Soft Computing Techniques for Cybersecurity.
M. A. Jabbar, Sanju Tiwari, Subhendu Kumar Pani, & Stephen Huang (Eds.)

prediction and diagnosis. These systems learn important features from biomedical datasets and make predictions based on new data. Similarly, the effectiveness of ML in predicting future outcomes has been recognized in industrial safety management. Proactive safety management requires the use of leading indicators to monitor and evaluate the effectiveness of protective measures, aiming to prevent incidents before they occur. In the agricultural sector, ML intelligent systems can assist farmers in optimizing crop yield by considering factors such as climate conditions, rainfall, soil conditions, irrigation systems, and pest and disease management. To address these challenges, this chapter proposes an intelligent approach for predicting accidents in industries, diagnosing diseases in medical settings, and predicting crop yield in agriculture. This chapter employs different ML algorithms and powerful classifiers to recognize potentially hazardous project characteristics and site conditions, predict diabetic retinal diseases, breast cancer, histopathology, tissue segmentation, cardiac diseases, chronic kidney diseases, and brain tumors at an early stage, and develop crop productivity as well as its quality in the agriculture sector. The system is further validated through statistical measures to determine the efficiency of the algorithms by comparing predictive accuracy and error rates.

6.1 INTRODUCTION

In recent days, the computer-aided decision support system has implemented machine learning (ML) approaches and has been practically applied in multidisciplinary fields. This system learns the important features of biomedical datasets and makes predictions based on new data. ML implementation now includes two-dimensional and three-dimensional images. ML-based algorithms are capable of predicting accidents and future safety measures to minimize accidents. However, these methods were developed for specific leading indicators, limiting their application as the impact on safety performance varies between organizations and projects. Research on predicting fire accidents aims to encourage the implementation of targeted security controls and improve safety performance. The main goal of precision agriculture is to improve crop yield. Crop growth and yield depend on environmental attributes such as climate variables, soil parameters, irrigation, pest, and disease management. ML algorithms and powerful classifiers have been developed to predict biomedical imaging and signaling, predict accidents, and predict crop productivity in the fields of medical, safety performance, and agriculture.

Diabetic retinopathy is one of the leading irreparable retinal diseases affecting vision [1]. Early-stage detection is crucial to prevent vision loss. The diagnosis of DR is achieved through the identification of exudates in fundus images. Classification prediction of DR is determined using ensemble ML algorithms such as AdaBoost, naïve bayes (NB), random forest (RF), and support vector machine (SVM) [2]. Deep learning (DL) convolutional neural networks (CNN), specifically GoogLeNet and AlexNet ImageNet models, are used for feature extraction with solvers like SGD, Adam, AdaGrad, NAG, and RMSprop to achieve a maximum accuracy of 74.5% [3]. Optical coherence tomography (OCT) images are utilized for effective classification of diabetic retinal diseases including Diabetic macular edema, DR, and age-related macular edema (AMD). The customized CNN OCT-NET, based on DL, achieves a precision of 93% and an Area Under Curve of 0.99 on the SD-OCT public dataset SERI-CUHK [4]. Diabetic macular edema affects the subretinal fluid (SRF) and the pigment epithelium detachment (PME) of the retina. A deep neural network combined with atrous spatial pyramid pooling is used for OCT image segmentation in DME [5], while a convolutional network is employed for exudates-based DME detection [6]. ML decision tree classifiers are applied for detecting diabetic macular edema lumps, achieving performance measures of 92% accuracy, 89.8% sensitivity, and 94.8% specificity [7].

The use of ML calculations requires a tremendous amount of data to trigger dynamic processes in various mechanical cycles. In this way, the implementation of innovative principles such as Cyber-physical systems and IoT enables the generation of different types of indigenous knowledge, as observed in the field of big data analytics. Generally, data follows a specific life cycle (source, collection, storage, processing, visualization, transmission, application). However, most of the time, the collected data that will be processed in subsequent steps is mixed with noisy data generated from the overall environment, making it difficult to separate the original dataset from the noise [8].

According to the statistical report by Liakos et al. [9], ML models are applied in various applications in agriculture such as crop yield prediction (20%), soil management (61%), and pest and disease detection (22%). In most cases, machine-based predictions obtained accurate forecasting in many aspects. It also proved that most agricultural studies implemented the artificial neural network model (ANN) and SVM model to make better decisions. Most probably, ANN models are used for implementation in crop improvement, soil, and water administration. When ML is implemented with sensor data, the models are evolving into artificial intelligence (AI) systems to provide valuable insights for making informed decisions with the definitive goal of enhancing crop productivity.

ML techniques can be utilized in the field of soil management to extract soil parameters. They can also be applied to predict crop growth based on factors such as soil moisture, weather conditions, and irrigation features. As a result, ML methods have contributed to efficient automatic irrigation [10]. Many farmers still follow traditional crop development systems, which may lead to decreased production in cases of insufficient rainfall. To enhance crop productivity, ML techniques with hybrid algorithms have been implemented, yielding promising results [11]. This chapter provides an overview of ML models used in various fields including medicine, safety management, and agriculture. The discussion focuses on how these models efficiently classify medical image data, analyze and predict fire accidents at an early stage, and identify and monitor crop canopy characteristics, soil humidity, pests, and diseases.

6.2 METHODS AND DISCUSSIONS

6.2.1 MACHINE LEARNING (ML) IN MEDICAL FIELD

6.2.1.1 DIABETIC RETINAL DISEASES DETECTION

Diabetic retinopathy is one of the leading irreparable retinal diseases affecting vision [12]. Detecting the early stage is crucial in preventing vision loss. The diagnosis of DR is achieved through the detection of exudates in the fundus image, as depicted in Figure 6.1. The classification and prediction of DR are determined using ensemble ML algorithms such as AdaBoost, NB, RF, and SVM [13].

OCT images are used for the effective classification of diabetic retinal diseases such as Diabetic macular edema, DR, and AMD (Figure 6.2). The DL customized CNN OCT-NET is used to achieve a precision of 93% and an Area Under Curve of 0.99 on the SD-OCT public dataset SERI-CUHK [14].

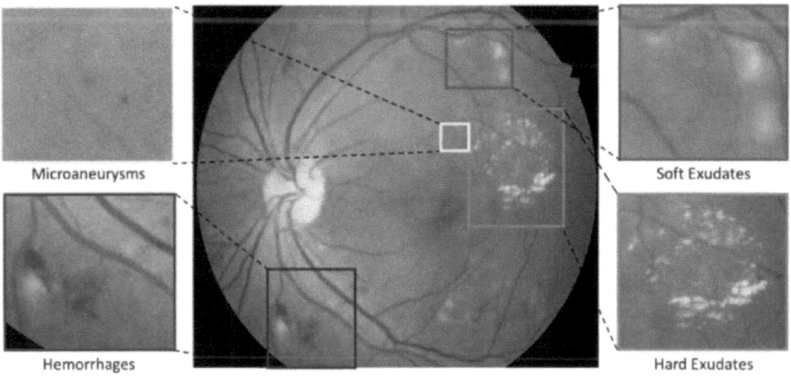

FIGURE 6.1 Diabetic retinopathy two-dimensional fundus image.

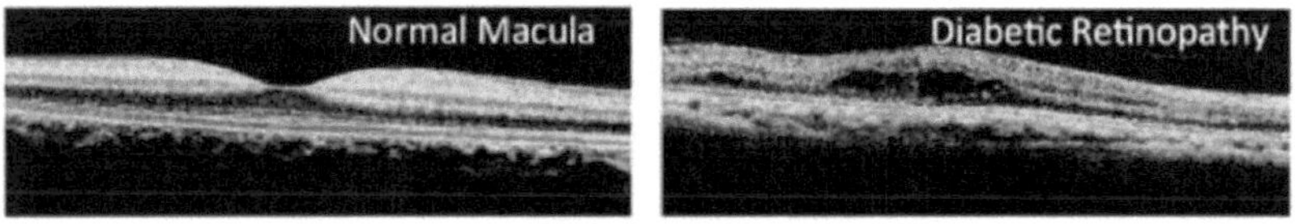

Diabetic Retinopathy OCT

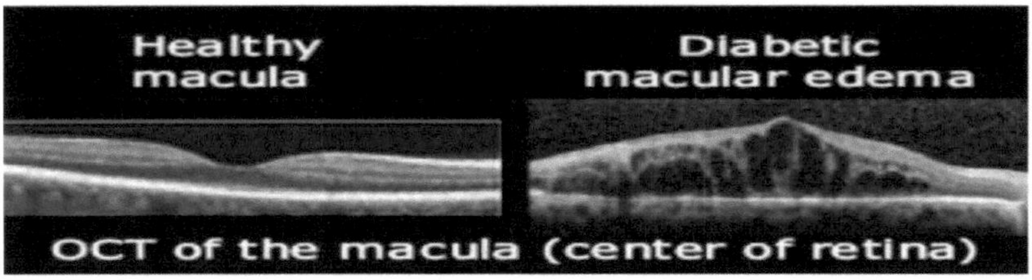

Diabetic Macular Edema

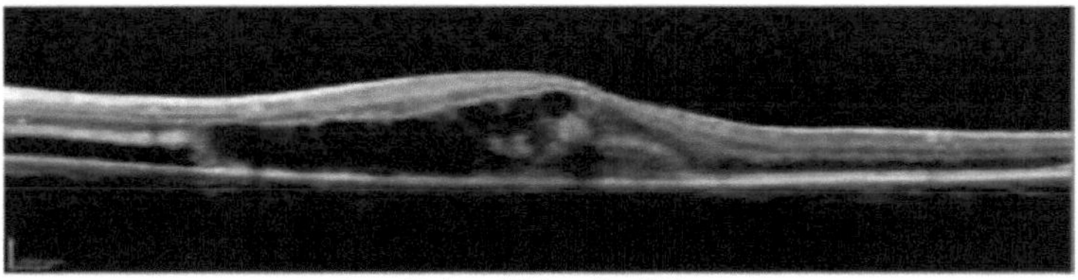

Age Related Macular Edema

FIGURE 6.2 OCT images of diabetic retinal diseases.

Diabetic macular edema affects the sub-retinal fluid (SRF) and PME of the retina. A deep neural network combined with atrous spatial pyramid pooling is used to segment the OCT image for DME [15], while a convolutional network is used for exudates-based DME detection. A ML decision tree classifier is applied to detect the diabetic macular edema lump, with performance measures of 92% accuracy, 89.8% sensitivity, and 94.8% specificity [16].

6.2.1.2 DIAGNOSIS OF BREAST CANCER

Breast cancer has been one of the common diseases affecting women worldwide. In 2018, approximately 15% of women died due to breast cancer, according to the World Health Organization (WHO) [17]. Early detection is crucial in order to prevent fatalities. ML algorithms have been utilized to detect breast cancer in mammogram images (Figure 6.3). A CNN, along with optimization methods, has been employed to classify the histopathological image dataset [18]. The performance of the classifier was evaluated based on the identification of malignant tumors, using techniques such as normalization, weight regularization, shallow SoftMax regression, and shallow logistic regression (LR) [19].

Multiple instance learning is used to identify histopathological images of breast cancer. A two-step DL model is employed for mammogram image classification. A deep convolutional pre-trained network is applied to achieve multi-view classification of mammogram images with an ROC of 0.9 [20].

Transfer learning techniques were implemented to achieve an AUC of 0.81 in the classification of breast cancer tumor mammographic images [21]. An extreme learning machine (ELM) technique was applied to diagnose the breast cancer mammographic dataset with an accuracy of 96.02% [20]. Tele mammographic images have played an important role in telehealth monitoring systems. ML was implemented using a J48 decision tree, RF, k-nearest neighbor (KNN), and multilayer perceptron (MLPs) [21].

6.2.1.3 BRAIN TUMOR DETECTION

Brain tumors are defined as the mass growth of abnormal cells in the brain, which can lead to death. Magnetic resonance imaging (MRI) is used for detecting this type of tumor (Figure 6.4). Nowadays, computer-aided systems are used for easy and fast diagnosis. CNN kernel-based

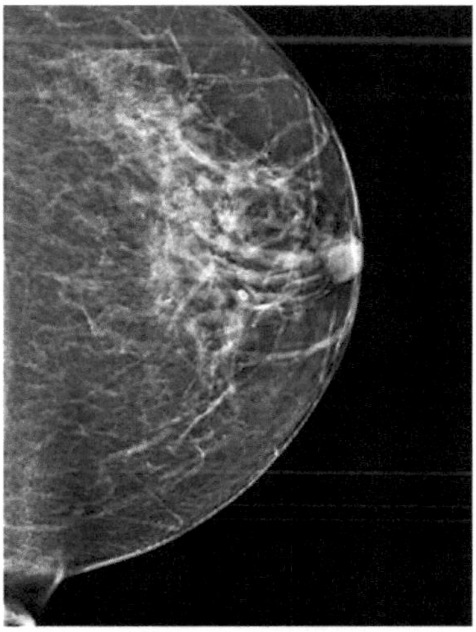

2D mammogram

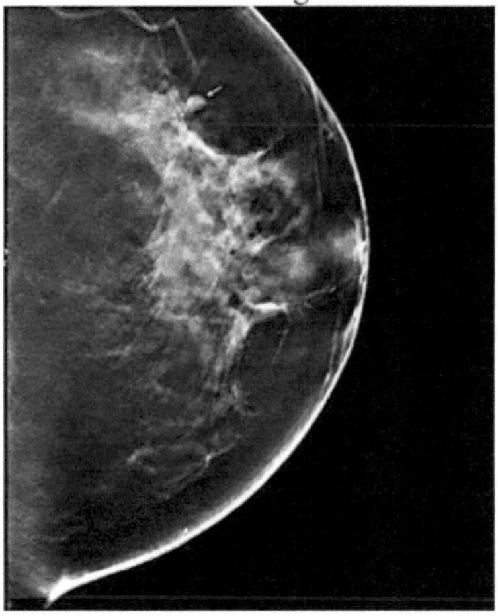

3D mammogram

FIGURE 6.3 Breast cancer mammogram images.

algorithms are implemented to segment the MRI brain images [22]. Deep neural network classifiers are applied to classify the types of brain tumors. The popular SVM ML algorithm is used for classifying brain MRIs. Auto multi-task medical image segmentation is demonstrated in the form of a single CNN that includes three tasks: brain MRI, Breast MRI, and cardiac CTA [23].

6.2.2 MACHINE LEARNING (ML) IN THE INDUSTRIAL SAFETY PERFORMANCE SECTOR

Humanity has achieved enormous advances in industrial manufacturing during the last 300 years. The third industrial revolution, which began in the 1950s, increased digitalization through the use of semiconductors and, more recently, communication networks, which made automated development possible. AI and ML have been extensively used in the manufacturing sector during the past 10 years, enabling more efficient operations, less resource waste and sustainable resource utilization, better working conditions, and improved quality and productivity. By allowing defect identification and prediction, optimum use of raw materials and resources, leveraging heterogeneous big data analysis, and networked manufacturing facilities, AI/ML-based manufacturing may bring about a number of industrial advancements [24].

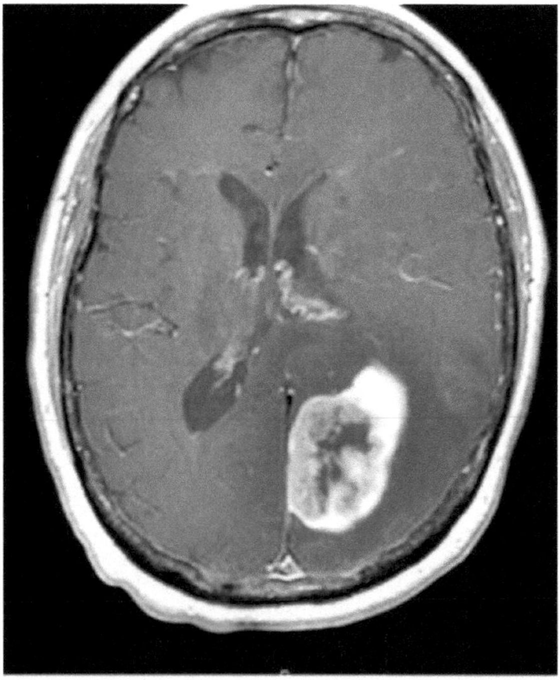

FIGURE 6.4 Brain tumor MRI image.

6.2.2.1 OCCUPATIONAL RISK PREDICTION

A branch of machine language is based on "deep learning" and demonstrates how some of the gaps in risk assessment can be addressed. ML refers to computer programming techniques for learning through experience. Through DL techniques, a computer can be trained to evaluate risk for safety-critical industries such as fireworks manufacturing, mining industries, and oil and gas. This involves analyzing a large volume of data, including indicators from regular operations and past incidents, ranging from mishaps to major accidents. Due to the subjective nature of the risk concept, professional oversight is required as the degree of risk cannot be applied with certainty to each case. This supervised learning enables DL. Once the model has been trained for risk categorization, it utilizes its expertise to assess major risks based on the condition of the controlled device, such as a fireworks, mining, or oil and gas field [25]. Figure 6.5 represents the development of DL for industrial risk assessment.

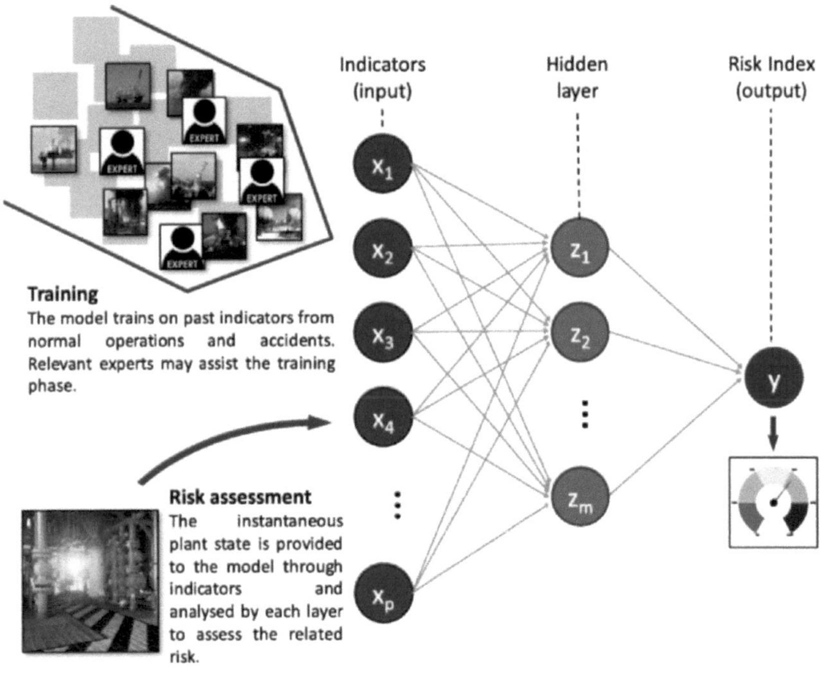

FIGURE 6.5 Industrial risk assessment approach by deep learning.

Source: Reprinted with permission from Ref. [52]https://creativecommons.org/licenses/by/4.0/

6.2.2.1.1 Classification and Regression (CART)

The classification and regression tree (CART) is a selected tree-based information extraction methodology used by numerous scientists for studying work-related accidents. The CART framework aims to establish a relationship between accident severity, working environmental factors, risk factors, and accident factors. By utilizing CART, it is possible to determine the anticipated scenarios and logical outcomes of business interactions within a real-world industry sector incidents [26].

6.2.2.1.2 Support Vector Machine (SVM)

Nonlinear kernel SVMs have been successfully and often used in supervised classification due to their capability to represent data in the required space where groups are clearly distinguishable. SVM algorithms have shown comparable or higher accuracy in regression and classification tasks compared to other empirical ML methods. SVM statistical models have very reliable output on binary classification issues [27].

The explanation for this is the SVM algorithm, which determines an optimum boundary capable of separating the two groups of the target variable. The three popular meta-heuristic algorithms for SVM include linear, radial basis function, and the degree d of the probability density function [28].

6.2.2.1.3 Decision Tree (DT)

Decision trees (DTs) have become more common because they are clear, easy to interpret, and efficient classification algorithms. The key reason for using DTs is to analyze quantity and quality trends in the data for the purpose of exploring hidden knowledge. Among the different classifications of DT algorithms, boosting has been considered one of the most significant developments in ML over the last 20 years, as it can transform a group of different learners [29].

6.2.2.1.4 Naïve Bayes (NB)

Bayesian classifiers are based on Bayes' theorem, which uses conditional probabilities to determine the likelihood of the dependent variable given the independent variables. The Naive Bayes classifier is one of the most widely used classifiers in AI. It simplifies the complexity of Bayesian classifiers by making the naive assumption of independence, reducing the number of parameters to be evaluated from 2^(2n–1) to just 2n when modeling P(X), where X is the independent variable, Y is the dependent variable, and n is the number of independent variables used in the experiment [30].

6.2.2.1.5 Reinforcement Learning (RL)

Specific reference to the unsupervised learning method involves determining if the chosen action resulted in a reward for a specific performance metric. Reinforcement learning (RL) looks at sequential actions and seeks out their outcomes, selecting those that are best suited to the problem at hand. Thus, RL is very different from other types of learning, which concentrate on the utilization of historical information, the production of knowledge from previous decisions and rewards [31]. The ML categories and key aspects are demonstrated in Figure 6.6.

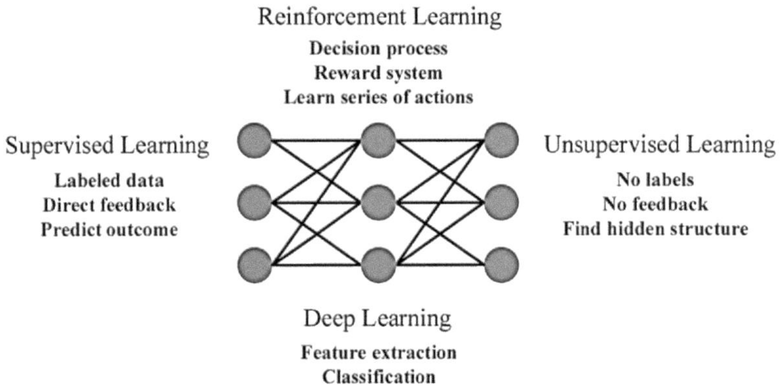

FIGURE 6.6 Machine learning categories and key aspects.

Source: Reprinted with permission from Ref. [53] https://creativecommons.org/licenses/by/

6.2.2.2 PREDICTIVE MODEL FOR SAFETY DOMAIN

Security leading indicators are an important component of proactive safety management and provide useful information on the performance of organizational protection [32]. Safety professionals have concluded that careful collection, evaluation, and mitigation of leading indicators would lead to real functional improvements [33]. Criteria have been established for selecting useful leading indicators, including validity, reliability, responsiveness, representativeness, openness to bias, and cost-effectiveness [34]. Figure 6.7 represents the key parameters identified in industrial safety and accidents.

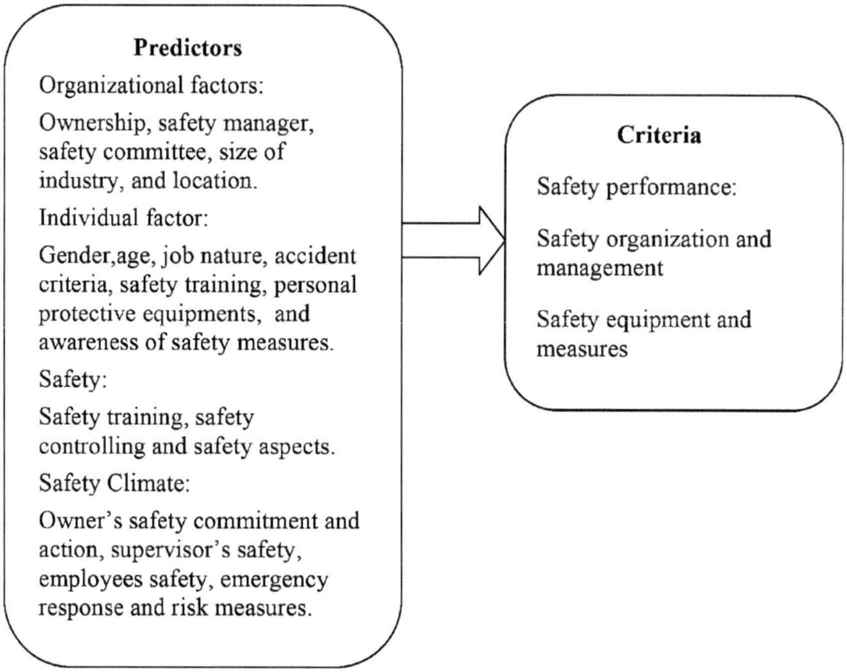

FIGURE 6.7 Occupational accidents risk and safety parameters.

Additionally, effective safety measures should be numerically assessed, updated frequently, and reflect the chosen overall safety determinants [35]. The potential impact of interaction related to the aggregation and analysis of multiple leading indicators is believed to yield the greatest anticipated outcome [34]. Previous studies suggest that promoting a zero-injury state and predictive protection models require more than just personal protective equipment (PPE) and proactive preventive strategies [37]. Performance may play a crucial role in bridging the gap. Companies will be able to develop accident prevention plans more swiftly and efficiently if they can anticipate and prepare for potential improvements in safety performance [38].

The capacity of a single company to lower accident rates via the use of cutting-edge and predictive analytics in conjunction with a rising safety data collection. To proactively anticipate where and when occupational injuries may occur, a variety of predictive models have been created [39]. Various models should be established to forecast construction safety performance globally. They have demonstrated that the diversity of claims and privilege instances in all built models, as well as the number of firms in every category of eligible profits, all predict the safety outcome. The prevalence of firms in various categories of responsible earnings and the frequency of workplace injuries were also shown to be positively and pertinently correlated [40].

To evaluate the precision of risk variables in predicting safety performance, employ a linear prediction model utilizing principal component analysis (PCA). To thoroughly analyze safety risk factors in infrastructure projects and ascertain the probability of a project's safety status, World Steel Association [41] proposed a hybrid system based on Bayesian networks. While their approach proved effective in forecasting safety outcomes, it was limited to small-scale projects.

6.2.2.3 PREDICTION OF OCCUPATIONAL ACCIDENTS

Occupational injuries are a significant concern across all sectors. According to statistics provided by the International Labor Organization (ILO), there are over 317 million reported incidents and hazardous events at work annually, resulting in more than 2.3 million premature deaths each year. In India alone, it is reported that over 48,000 employees die annually due to work-related accidents, and nearly 37 million cases result in at least 3 days of sickness absence. The study also reveals a fatal accident rate of 114 per 1,000,000 employees and an accident rate of 8,700 per 1,000,000 employees in India [38].

As per the World Steel Association, the protection and health of workers are important for sustainability [42]. Occupational risk is defined by two parameters: one that shows the incidence of injuries and the other that relates to the severity of injuries. Occupational accident reports can be classified in a format that includes parameters such as the year of the accident, place, workers' criteria, age of the workers, time of the incidents, description of the accidents, etc. The above criteria include the type of injury, the substance factor involved, the types of injury, the part of the body impacted, the hazardous condition involved in the incident, and the hazardous activity involved in the incident.

The optimization approach is to create a measurement model that uses the GA methodology to determine occupational hazards, taking into account the magnitude and frequency of occupational injuries. It is also reported that safety in the industry is correlated with higher performance, and the most effective steelmakers are also found to be the safest. Therefore, it is a major concern today that, while we have sufficient experience of severe, harmful, and time-loss incidents, we may not learn from past records. If we could predict future instances of occurrences using historical evidence, this would help the concerned industry. Forecasts have also been produced in the field of workplace injury analysis.

It has been described as an essential strategy that is focused on the fundamental premise that incidents can be avoided if they can be expected. As a consequence, the prediction of workplace injuries in the practice of safety analysis at work has been shown to be important in order to minimize the risk of injuries that could, in turn, occur [43].

6.2.2.4 OCCUPATIONAL MODELS

Identify the new technologies that could be associated to ML [44]:

1. **Predictive Maintenance:** The ability to accurately predict challenges to the manufacturing line prior to an interruption is invaluable to the manufacturer. It enables the employer to plan downtime during the most suitable times and reduce unnecessary delays. Unscheduled interruptions heavily impact gross profit and can result in the loss of the consumer base. The implementation of ML for predictive maintenance is expected to increase by around 38% across factories, as it has the potential to improve gross profit by eliminating unscheduled shutdowns.
2. **Personal Protective Equipment (PPE):** It is a device that protects individuals from health and safety hazards. Failing to wear PPE significantly increases the risk of injury and, in some cases, can result in financial penalties for injuries or even the loss of life for staff members who do not wear protective gear, such as face masks and safety glasses. The implementation of ML for PPE

identification allows companies to minimize the risk of worker injuries, prevent accidents in the workplace, improve operational efficiency, and gather data on workplace safety.

3. **Smart Manufacturing:** A key objective of AI and AI is to enable the development of a digital replica of the manufacturing floor. The creation of the digital twin would occur through an effort under a model-based systems engineering process (MBSE) using AI algorithms and acquired data as a foundation. The digital twin would serve as a platform for conducting "what if" scenarios to uncover unknowns. Additionally, the digital twin can be used as an end-to-end model for designing higher reliability components and optimizing the interactions between production line machines to enhance performance [45].
4. **Machine Human Interaction:** The division of labor between humans and machines is typically observed in technologically advanced operational settings, and the jobs assigned are classified as machine- and human-specific. Unlike the concept of computer-integrated manufacturing (CIM), Industry 4.0 aims to combine people and machines in a collaborative workspace within a human-cyber-physical framework (HCPS), as opposed to a worker-less production facility [26].
5. **Open Access:** The most recent advancements in ML-driven techniques and the emergence of DL have opened up new opportunities to revolutionize Industry 4.0 and beyond. These advancements have led to the possibility of achieving highly autonomous and efficient industrial operations, while improving fault detection, mitigating human errors, ensuring security, and preventing unexpected losses, even in complex and dynamic environments [43].

The integration of AI and ML in various aspects of human society has played a crucial role in enabling exciting new opportunities to enhance quality of life and preserve the sustainability of our ecosystem. In the context of Industry 4.0, AI/ML enables numerous improvements across the manufacturing sector, enhancing processes such as waste reduction, automation, and product quality. A highly significant area in realizing the vision of Industry 4.0 is the timely detection and prevention of faults, utilizing solutions from the industrial internet of things, and the subsequent acquisition and processing of diverse large-scale data [46]. Figure 6.8 illustrates the use of ML algorithms for industrial risk assessment prediction and prevention.

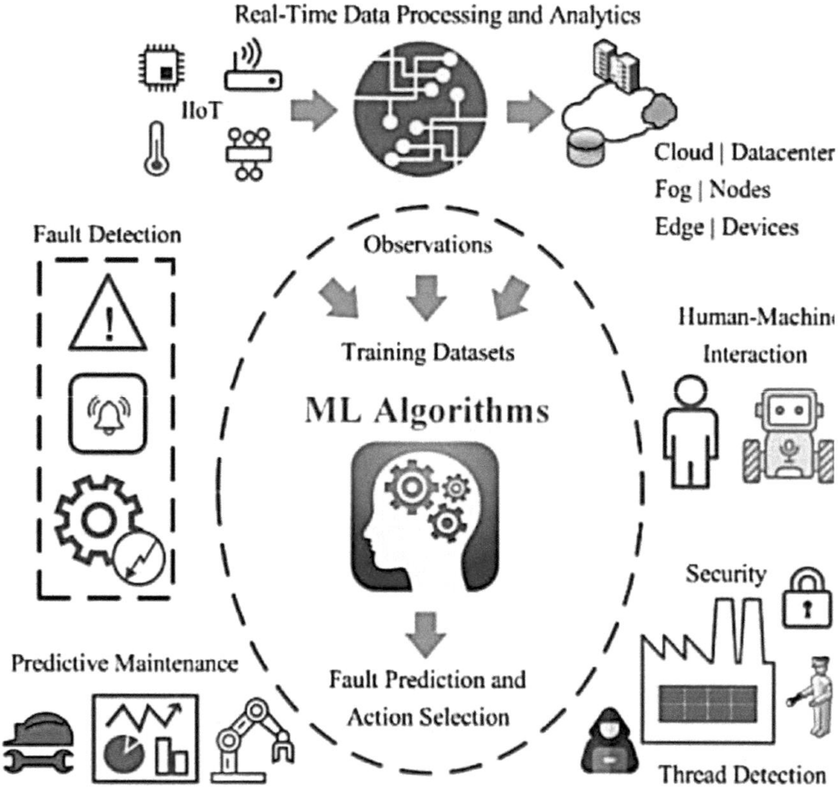

FIGURE 6.8 Industrial risk assessment prediction and prevention using ML.

Source: Reprinted with permission from Ref. [53] https://creativecommons.org/licenses/by/

6.3 MACHINE LEARNING (ML) IN AGRI-TECHNOLOGY DOMAIN

6.3.1 WEATHER AND CROP PREDICTION

The yield of crops is primarily based on weather conditions and soil variables such as pH value and nutrients. Weather parameters like temperature, solar radiation, relative humidity, and rainfall play a significant role in the growth and development of crops. Forecasting crop yield is challenging due to the fluctuation of weather variables in meteorological data.

Recurrent neural networks (RNNs) can be effectively used in weather forecasting [47]. Additionally, crops can be classified based on weather using classification algorithms such as SVM, C4.5, C5.0, CART, and RF, as shown in Figure 6.9. Although RNN successfully processes time series data, integrating RNN with a statistical model like generalized autoregressive conditional heteroskedasticity (GARCH) enhances the prediction process. Compared to conventional neural networks, RNN allows the output from the previous step to be forwarded to the current step, enabling the hidden layers in RNN to remember the sequence of the network. Therefore, the values are associated with each other [48]. Before classifying crops or forecasting weather, data preprocessing is necessary to fill in missing values in the climatic or crop data.

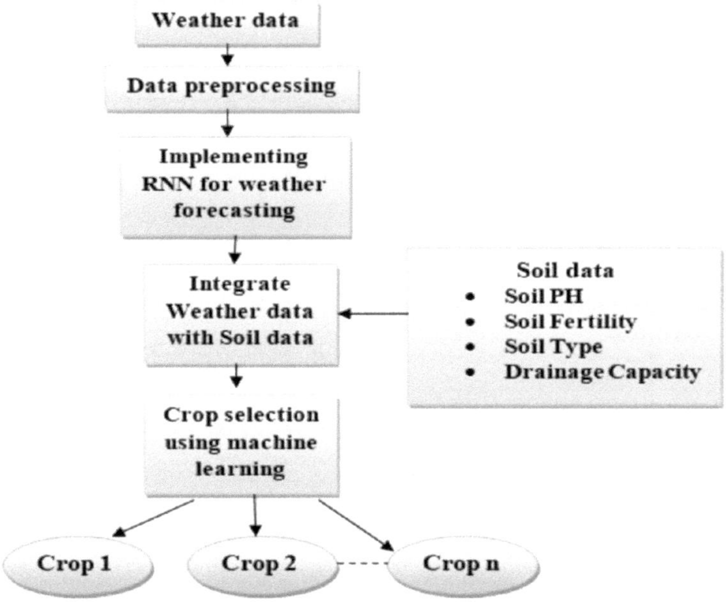

FIGURE 6.9 Weather data for crop prediction.

6.3.2 *PEST AND DISEASE MANAGEMENT*

Crops must be protected from pests and diseases to enhance crop yield through efficient ML methods. ML decision tree algorithms like C4.0 and C5.0 can be used to classify agricultural crop pest images. Both algorithms construct a decision tree using a training set of data [49]. Accuracy prediction is done using a split testing set of data. C4.5 and C5.0 are both effective classifiers for constructing DT and generating sets of rules. The C4.5 algorithm uses a depth-first strategy during construction, allowing for pruning and compact handling of missing and noisy data. However, C5.0 outperforms C4.5 in terms of processing speed, memory, and efficiency in some cases [50]. CNNs are a type of neural network that can be applied to classify pest images in the agricultural field. Feature extraction in pest images is achieved through adaptive learning of neurons in the neural network [51]. The flow chart for pest recognition and classification is schematically represented in Figure 6.10.

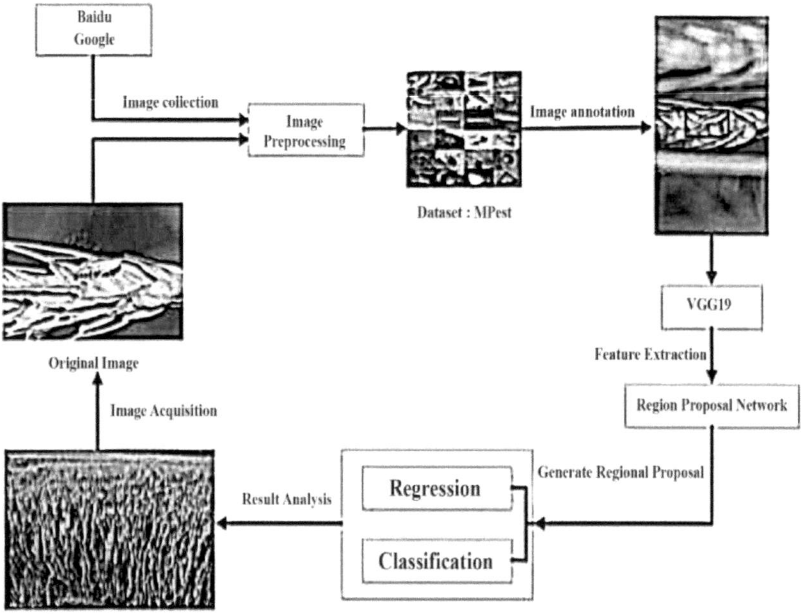

FIGURE 6.10 Flow chart of pest recognition and classification.

6.3.3 SOIL AND IRRIGATION MANAGEMENT

All crops grow better depending on the type of soil, nutrients, and irrigation. Therefore, it is important to choose the appropriate soil for planting in order to maximize crop yield. Changes in soil and weather conditions have a definite impact on the water management system, leading to a decrease in crop yield. Therefore, it is recommended to use intelligent automatic irrigation systems that utilize soil sensor data and weather data, as shown in Figure 6.11. The soil moisture level data is transmitted to the cloud via a Wi-Fi modem. ML classification techniques are employed to classify irrigation strategies based on the soil data. As a result, water usage by automatic irrigation systems can be reduced while achieving a high accuracy model [36].

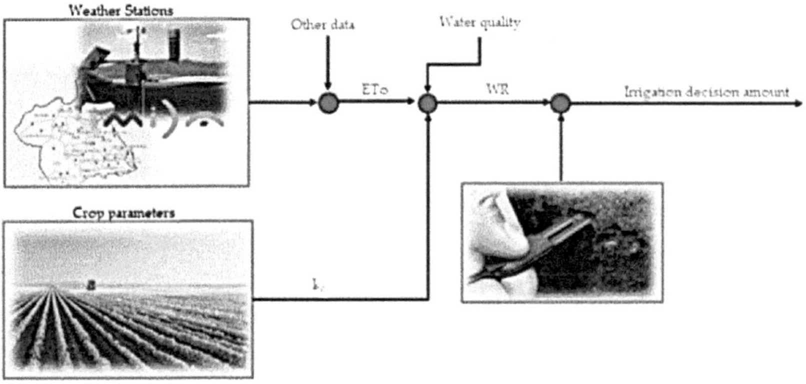

FIGURE 6.11 Intelligent automatic irrigation model.

6.4 CONCLUSION

Machine learning (ML) models have advanced rapidly over the past few years. The development of these algorithms is likely to provide more meaningful datasets, which, when integrated with complex algorithmic results, pave the way for better prediction and decision-making in the fields of medicine, safety management, and agriculture. Numerous ML models have already been intelligently implemented in various domains. This chapter predominantly reveals that:

- CNN and other DL models with optimization offer an effective image classification in detecting cancer and tumors.
- Classification, regression, and network models are appropriate for predicting fire accidents in the cracker's city at the earliest stage. These models are considered promising tools for conducting regression analysis of risk assessment in industries, enabling the

measurement of accidents, and providing valuable knowledge for safety management measures.

- Machine learning models such as SVM, Decision tree, NB, and neural networks enhance the prediction of crop yield, soil management, irrigation systems, pests, and disease management.

ACKNOWLEDGMENTS

The first three authors would like to express their gratitude to the management of Kalasalingam Academy of Research and Education for providing the fellowship to carry out the research work.

KEYWORDS

- **agri-technology**
- **disease management**
- **intelligent model**
- **machine learning algorithm**
- **medical field**
- **safety management**

REFERENCES

1. Nathan, G. C., David, S. F., & Thomas, L., (2003). Important causes of visual impairment in the world today | *JAMA* | JAMA Network. https://jamanetwork.com/journals/jama/article-abstract/197460 (accessed on 20 February 2024).
2. Bhatia, K., Arora, S., & Tomar, R., (2016). Diagnosis of diabetic retinopathy using machine learning classification algorithm. In: *2016 2nd International Conference on Next Generation Computing Technologies (NGCT)* (pp. 347–351). doi: 10.1109/NGCT.2016.7877439.
3. Lam, C., Yi, D., Guo, M., & Lindsey, T., (2018). Automated detection of diabetic retinopathy using deep learning. *AMIA Jt Summits Transl. Sci. Proc., 2018*, 147–155.
4. Perdomo, O., Rios, H., Rodeiguez, F. J., Otalora, S., Meriaudeau, F., Muller, H., & Gonazalez, F. A., (2019). Classification of diabetes-related retinal diseases using a

deep learning approach in optical coherence tomography. *Computer Methods and Programs in Biomedicine, 178*, 181–189. doi: 10.1016/j.cmpb.2019.06.016.
5. Junjie, H., Yuanyuan, C., & Zhang, Y., (2019). Automated segmentation of macular edema in OCT using deep neural networks. *Medical Image Analysis, 55*, 216–227.
6. Juan, M., Lei, Z., & Yangqin, F., (2018). Exudate-based diabetic macular edema recognition in retinal images using cascaded deep residual networks. *Neurocomputing, 290*, 161–171. Elsevier.
7. Mahendran, G., Periyasamy, M., Murugeswari, S., & Karthika, D. N., (2020). Analysis of retinal diseases using machine learning algorithms. *Materials Today: Proceedings*. doi: 10.1016/j.matpr.2020.03.691.
8. Rehman, M. H., Yaqoob, I., Salah, K., Imar, M., Jayaraman, P. P., & Perera, C., (2018). The role of big data analytics in industrial internet of things. *Future Gener. Comput. Syst., 99*, 247–259.
9. Liakos, K. G., Busato, P., Moshou, D., Pearson, S., & Bochtis, D., (2018). Machine learning in agriculture: A review. *Sensors, 18*(8), 2674.
10. Storm, H., Baylis, K., & Heckelei, T., (2020). Machine learning in agricultural and applied economics. *European Review of Agricultural Economics, 47*(3), 849–892.
11. Bhimanpallewar, R. N., & Narasingarao, M. R., (2020). Alternative approaches of machine learning for agriculture advisory system. In: *2020 10th International Conference on Cloud Computing, Data Science & Engineering (Confluence)* (pp. 27–31). IEEE.
12. WHO | *Breast Cancer*. WHO. http://www.who.int/cancer/prevention/diagnosis-screening/breast-cancer/en/ (accessed on 20 February 2024).
13. Kwadwo, S., You, F., & Tettey, O., (2020). Efficient, ultra-facile breast cancer histopathological images classification approach utilizing deep learning optimizers. *IJCA, 177*(37), 1–9. doi: 10.5120/ijca2020919875.
14. Ferdaus, J., (2020). *Diagnosis of Breast Cancer Using Neural Networks with Different Logistic Regressions*. [Online]. Available: https://www.doria.fi/handle/10024/177454 (accessed on 20 February 2024).
15. Sudharshan, P. J., Petitjean, C., Spanhol, F., Oliveira, L. E., Heutte, L., & Honeine, P., (2019). Multiple instance learning for histopathological breast cancer image classification. *Expert Systems with Applications, 117*, 103–111. doi: 10.1016/j.eswa.2018.09.049.
16. Dhungel, N., Carneiro, G., & Bradley, A. P., (2016). The automated learning of deep features for breast mass classification from mammograms. In: *Medical Image Computing and Computer-Assisted Intervention – MICCAI 2016* (pp. 106–114). Cham doi: 10.1007/978-3-319-46723-8_13.
17. Carneiro, G., Nascimento, J., & Bradley, A. P., (2015). Unregistered multiview mammogram analysis with pre-trained deep learning models. In: *Medical Image Computing and Computer-Assisted Intervention – MICCAI 2015* (pp. 652–660). Cham. doi: 10.1007/978-3-319-24574-4_78.
18. Huynh, B. Q., Li, H., & Giger, M. L., (2016). Digital mammographic tumor classification using transfer learning from deep convolutional neural networks. *JMI, 3*(3), 034501. doi: 10.1117/1.JMI.3.3.034501.

19. Xie, W., Li, Y., & Ma, Y., (2016). Breast mass classification in digital mammography based on extreme learning machine. *Neurocomputing, 173*, 930–941. doi: 10.1016/j.neucom.2015.08.048.
20. Syed, L., Jabeen, S., & Manimala, S., (2018). Telemammography: A novel approach for early detection of breast cancer through wavelets-based image processing and machine learning techniques. In: Hassanien, A. E., & Oliva, D. A., (eds.), *Advances in Soft Computing and Machine Learning in Image Processing* (pp. 149–183). Cham: Springer International Publishing.
21. Pereira, S., Pinto, A., Alves, V., & Silva, C. A., (2016). Brain tumor segmentation using convolutional neural networks in MRI images. *IEEE Transactions on Medical Imaging* (Vol. 35, No. 5, pp. 1240–1251). doi: 10.1109/TMI.2016.2538465.
22. Mohsen, H., El-Dahshan, E. S. A., El-Horbaty, E. S. M., & Salem, A. B. M., (2018). Classification using deep learning neural networks for brain tumors. *Future Computing and Informatics Journal, 3*(1), 68–71. doi: 10.1016/j.fcij.2017.12.001.
23. Gurusamy, R., & Subramaniam, D. V., (2017). *A Machine Learning Approach for MRI Brain Tumor Classification,* 18.
24. Lu, J., (2017). Industry 4.0: A survey on technologies, applications, and open research issues. *J. Ind. Inf. Integr., 6*, 1–10.
25. Goodfellow, I. J., Bengio, Y., & Courville, A., (2016). *Deep Learning*. The MIT Press, Citeseer, Cambridge, Massachusetts, US.
26. Cheng, C., Leu, S., Cheng, Y., Wu, T., & Lin, C., (2012). Applying data mining techniques to explore factors contributing to occupational injuries in Taiwan's construction industry. *Accident Analysis & Prevention, 48*, 214–222.
27. Xu, L. D., & Duan, L., (2019). Big data for cyber-physical systems in Industry 4.0: A survey. *Int. J. Enterp. Inf. Syst., 13*, 148–169.
28. Reiman, T., & Pietikäinen, E., (2011). Leading indicators of system safety: Monitoring and driving the organizational safety potential. *Safety Science, 50*(10), 1993–2000.
29. Indumathi, N., Ramalakshmi, R., & Ajith, V., (2021). Analysis of risk factors in the firework industries: Using decision tree classifier. In: *2021 International Conference on Advance Computing and Innovative Technologies in Engineering (ICACITE)* (pp. 811–814). IEEE.
30. Hale, A., (2009). Why safety performance indicators? *Safety Science, 47*(4), 479, 480.
31. Øien, K., Utne, I. B., & Herrera, I. A., (2011). Building safety indicators: Part 1 – Theoretical foundation. *Safety Science, 49*(2), 148–161.
32. Hinze, J., Thurman, S., & Wehle, A., (2013). Leading indicators of construction safety performance. *Safety Science, 51*(1), 23–28.
33. Schultz, G. *Using Advanced Analytics to Predict and Prevent Workplace Injuries*. Online: https://ohsonline.com/Articles/2012/07/01/UsingAdvanced-Analytics-to-Predict-and-PreventWorkplace-Injuries.aspx (accessed on 20 February 2024).
34. Ghodrati, N., Yiu, T. W., Wilkinson, S., & Shahbazpour, M., (2018). A new approach to predict safety outcomes in the construction industry. *Safety Science, 109*, 86–94.
35. Kuhn, M., & Johnson, K., (2013). *Applied Predictive Modeling*. Springer, New York, NY.
36. Torres-Sanchez, R., Navarro-Hellin, H., Guillamon-Frutos, A., San-Segundo, R., Ruiz-Abellón, M. C., & Domingo-Miguel, R., (2020). A decision support system for

irrigation management: Analysis and implementation of different learning techniques. *Water, 12*(2), 548.
37. Esmaeili, B., Hallowell, M. R., & Rajagopalan, B., (2015). Attribute-based safety risk assessment. II: Predicting safety outcomes using generalized linear models. *Journal of Construction Engineering and Management, 141*(8), 04015022.
38. Xia, N., Zou, P. X. W., Liu, X., Wang, X., & Zhu, R., (2018). A hybrid BN-HFACS model for predicting safety performance in construction projects. *Safety Science, 101*, 332–343.
39. *Safety and Health at Work*, (2016). Ilo.org, [Online]. Available: http://www.ilo.org/global/topics/safety-and-healthat-work/lang--en/index.htm (accessed on 20 February 2024).
40. Hämäläinen, P., Takala, J., & Saarela, K., (2006). Global estimates of occupational accidents. *Safety Science, 44*(2), 137–156.
41. World Steel Association – Safety and Health, (2016). Worldsteel.org, [Online]. Available: http://www.worldsteel.org/steel-by-topic/safety-and-health.html (accessed on 20 February 2024).
42. https://www.machinedesign.com/automation-iiot/article/21838038/how-machine-learning-is-transforming-industrial-production (accessed on 20 February 2024).
43. Hastie, T., Tibshirani, R., & Friedman, J., (2009). *The Elements of Statistical Learning: Data Mining, Inference, and Prediction.* Springer Science & Business Media.
44. Zhou, J., Li, P., Zhou, Y., Wang, B., Zang, J., & Meng, L., (2018). Toward new-generation intelligent manufacturing. *Engineering, 4*, 11–20.
45. Zheng, X., Wang, M., & Ordieres-Meré, J., (2018). Comparison of data preprocessing approaches for applying deep learning to human activity recognition in the context of Industry 4.0. *Sensors, 18*, 2146.
46. Lesjak, C., Hein, D., & Winter, J., (2015). Hardware-security technologies for industrial IoT: TrustZone and security controller. In: *Proceedings of the IECON 2015—41st Annual Conference of the IEEE Industrial Electronics Society* (pp. 2589–2595). Yokohama, Japan.
47. Jain, S., & Ramesh, D., (2020). Machine Learning convergence for weather-based crop selection. In: *2020 IEEE International Students' Conference on Electrical, Electronics and Computer Science (SCEECS)* (pp. 1–6). IEEE.
48. Murali, P., Revathy, R., Balamurali, S., et al., (2020). Integration of RNN with GARCH refined by whale optimization algorithm for yield forecasting: A hybrid machine learning approach. *J. Ambient Intell. Human Comput.* https://doi.org/10.1007/s12652-020-01922-2.
49. Revathy, R., & Lawrance, R., (2017). Classifying crop pest data using C4. 5 algorithm. In: *2017 IEEE International Conference on Intelligent Techniques in Control, Optimization and Signal Processing (INCOS)* (pp. 1–6). IEEE.
50. Revathy, R., Balamurali, S., & Lawrance, R., (2019). Classifying agricultural crop pest data using Hadoop MapReduce based C5. 0 Algorithm. *Journal of Cyber Security and Mobility, 8*(3), 393–408.
51. Xia, D., Chen, P., Wang, B., Zhang, J., & Xie, C., (2018). Insect detection and classification based on an improved convolutional neural network. *Sensors, 18*(12), 4169.

52. Nicola Paltrinieri, Louise Comfort, Genserik Reniers (2019). Learning about risk: Machine learning for risk assessment, Safety Science, 118, 475-486,
53. Angelopoulos A, Michailidis ET, Nomikos N, Trakadas P, Hatziefremidis A, Voliotis S, Zahariadis T. Tackling Faults in the Industry 4.0 Era—A Survey of Machine-Learning Solutions and Key Aspects. Sensors. 2020; 20(1):109. https://doi.org/10.3390/s20010109

CHAPTER 7

THE USAGE OF CLOUD COMPUTING AND IOT IN ARTIFICIAL INTELLIGENCE

POONAM SINHA[1] and SUNITA GOND[2]

[1]*Head, Department of Electronics and Communication Engineering, BUIT, Barkatullah University, Bhopal, Madhya Pradesh, India*

[2]*Faculty, Department of Master of Computer Application, BUIT, Barkatullah University, Bhopal, Madhya Pradesh, India*

ABSTRACT

With the accelerating development and application of cloud computing and computer science technology, IoT, and cloud computing play an important role in various applications such as medical-assisted environments, agriculture fields, and AI cloud, among others. The concept of combining artificial intelligence (AI) with cloud computing is currently prevailing and being implemented by enterprises. An AI cloud essentially consists of shared infrastructure for AI, supporting numerous projects and various workloads simultaneously, along with the internet of things (IoT). The IoT is the original application for cloud computing, as both cloud computing and IoT are two distinct technologies with applications in our day-to-day lives. A novel model that merges both cloud computing and IoT is predicted to emerge in the future, providing an enhanced and efficient platform.

The cloud providers are joining IoT because of the innovation it offers, providing a ready-made infrastructure where devices can seamlessly

The Fusion of Artificial Intelligence and Soft Computing Techniques for Cybersecurity.
M. A. Jabbar, Sanju Tiwari, Subhendu Kumar Pani, & Stephen Huang (Eds.)

connect and utilize its services. Collaboration between cloud providers and device makers enables access to advanced technology and monitoring, fostering constant innovation and delivering a reliable and seamless flow of data to consumers.

IoT is not just about devices communicating with us through AI, but also about devices communicating with each other. In the IoT, cloud applications act as mediators and communication assistants. With powerful APIs (Application Platform Interfaces) available for application developers and product creators, cloud communications function as plug-and-play solutions, similar to those used in our smartphones.

This mobility revolution prompts us to consider privacy and security concerns associated with the combination of IoT and cloud computing. However, the cloud computing infrastructure and the latest APIs in IoT can address these issues by enhancing technology and providing advantages such as remote processing power, easy monitoring, smooth inter-service, and inter-device communications, as well as ensuring user security and privacy.

7.1 CLOUD COMPUTING: INTRODUCTION AND APPLICATIONS

Cloud computing is a model of computer networking that provides access to various computer system resources, particularly data storage and computing power, without direct user interference and management. Data is available at all times for multiple users through internet-connected data centers. Large-scale clouds are made available to users at multiple locations, connected to central servers for proper functioning of the cloud.

Cloud data and environments can be made private, limited to a specific organization, or made accessible to all cloud users. Cloud computing is based on the principle of resource sharing, achieving connectivity and cost-effectiveness.

Migrating to a cloud environment allows companies to reduce or eliminate costs associated with investing in basic IT infrastructure. It offers various benefits, including fast application performance during peak demand, enhanced manageability, minimal maintenance, high computational resources, and flexibility in hardware and software.

It is a pay-as-per-user model, where users only pay for the resources, they are using and availing from cloud providers. Cloud providers offer on-demand services to the users and maintain quality of service (QoS) and various security measures at their own cost and risk (Figure 7.1).

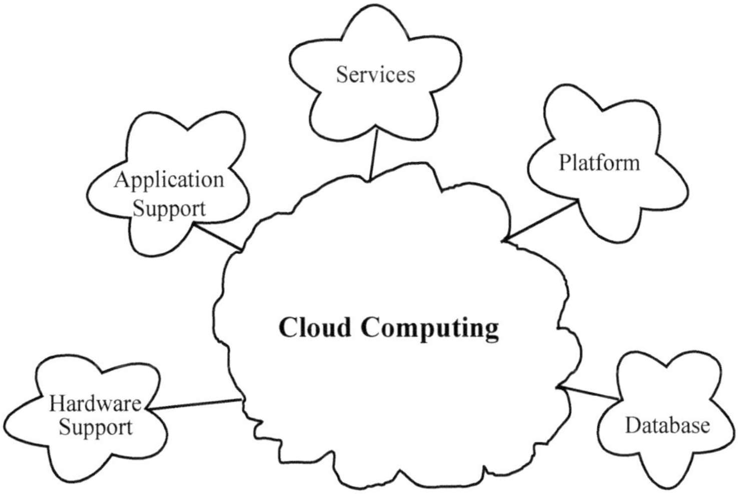

FIGURE 7.1 Cloud computing.

7.1.1 MERITS OF CLOUD COMPUTING

1. **Scalable:** Cloud services provide quick scaling up and down of computing resources to meet the various demands of users.
2. **Affordable:** Cloud services offer lower costs as hardware upgrades and maintenance responsibilities are handled by the service provider.
3. **Secure:** Service providers offer various secure cloud services to ensure the safety and security of our data in a third-party environment, maintaining strict security protocols.

7.2 VARIOUS MODELS OF CLOUD COMPUTING

Three types of models supported by cloud computing are:

- SaaS (software-as-a-service);
- IaaS (infrastructure-as-a-service); and
- PaaS (platform-as-a-service).

Each of the cloud models has its own benefits to serve the needs of various businesses. We need to understand the requirements and benefits according to our specific needs (Figure 7.2).

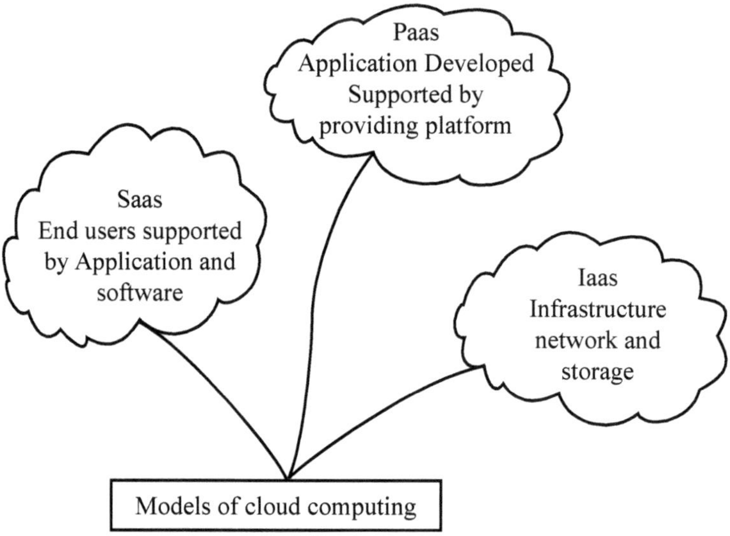

FIGURE 7.2 Cloud services model.

1. **Software-as-a-Service (SaaS):** It is a model that provides a fast approach to various cloud-based web applications. The service providers control the overall computing stack, which can be easily accessed through any web browser. Various applications that run on the cloud can be used on a paid basis, while some are also available for free access with limited features.

   In the pre-existing computing infrastructure of your organization, SaaS does not require any type of installations or downloads, as application maintenance and installation are all handled by the service providers. Some well-known examples of SaaS are Google G Suite and Drop box (Figure 7.3).

2. **Platform-as-a-Service (PaaS):** It provides us with a platform to develop, test, and organize different applications according to our business needs. Implementing PaaS transforms the process of enterprise software development by offering a virtual runtime environment that allows us to develop and test applications in a flexible and convenient manner. Service providers also offer a flexible pricing scale, allowing us to choose resources such as servers, storage, and networking based on our requirements. These resources can be managed either by the company itself or

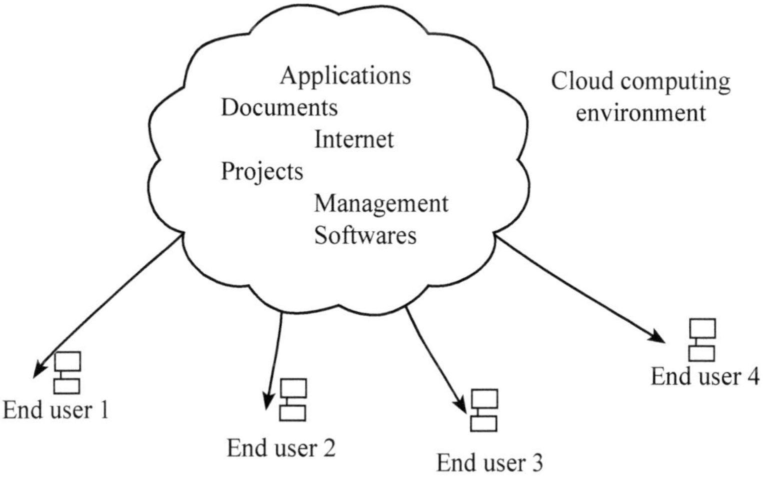

FIGURE 7.3 Software-as-a-service (SaaS).

by a service platform provider. Some important and well-known examples include Google App Engine and AWS Elastic Beanstalk (Figure 7.4).

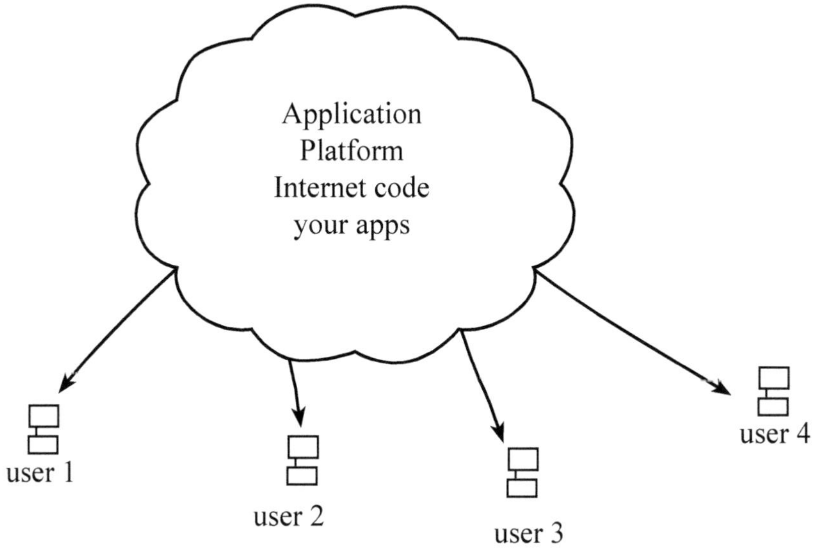

FIGURE 7.4 Platform-as-a-service (PaaS).

3. **Infrastructure-as-a-Service (IaaS):** Virtualization of various computing resources over the cloud is provided to us by service providers. These services are called IaaS. An integral range of computing infrastructures, such as storage, servers, and networking hardware, along with maintenance and support, are provided to us by IaaS cloud providers. Businesses can prefer computing resources on required premises without the need for hardware installation. Some important and well-known examples include Amazon Web Services and Microsoft Azure (Figure 7.5).

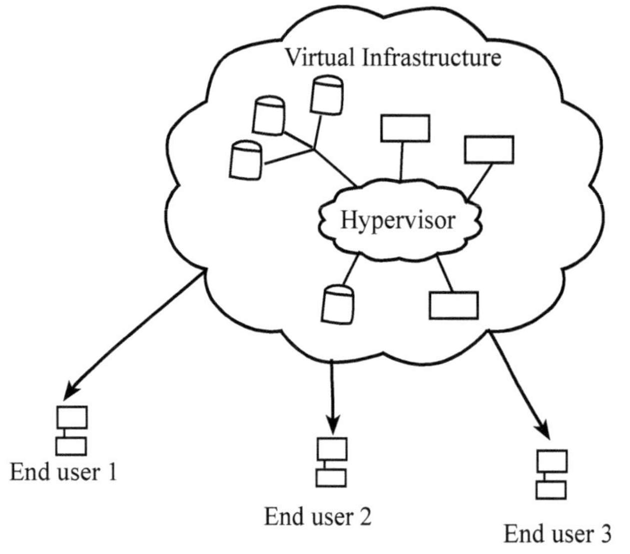

FIGURE 7.5 Infrastructure-as-a-service (IaaS).

7.2.1 BENEFITS OF SAAS

1. **Less Investment:** SaaS reduces the costs involved in the purchase of infrastructure, installation charges, maintenance, and upgrade costs of computing hardware.
2. **Accessibility:** SaaS provides us with an environment that allows us to access services from anywhere, using any device such as smartphones. This is helpful in eliminating the constraints associated with on-premises software.

3. **Prepared Environment:** We can easily and quickly use SaaS services, which means they become functional in a fraction of the time. You only need to sign up to avail the services and access fast and powerful computing resources provided by the environment.

SaaS provides services based on communication, content delivery, and simplified scheduling of meetings. It is best suited for small-scale businesses and industries that do not have the additional budget and resources to invest in on-premises hardware.

7.2.2 BENEFITS OF IAAS

1. **Reduces Costs:** This cloud model eliminates the need for on-premises hardware, thus reducing the costs associated with hardware infrastructure.
2. **Scalability Increases:** It is the most flexible cloud computing model that allows us to scale the computing resources up or down as per our demand.
3. **Simple Structure:** This model easily deploys the services, including servers, processing, storage, and networking, to ensure their availability at all times.

IaaS is the most flexible cloud model, giving us the best option when it comes to IT hardware infrastructure. However, it is costlier than SaaS or PaaS cloud models.

7.2.3 BENEFITS OF PAAS

1. **Minimize Time of Development:** This cloud model can reduce the development time as it provides various computing resources, such as server-side components, which simplify the process. This allows the development team to focus solely on the development part.
2. **Support Various Programming Language:** It offers a platform to support multiple programming languages, which is useful for a software development company that can easily utilize the environment to build applications for various projects.

3. **Support Team Collaboration:** Whether team members of a project can live anywhere, this cloud model can enhance collaboration between team members through a common platform. With multiple developers and multiple service providers, this cloud model works efficiently. It supports the creation of customized applications by providing a process of application development that minimizes total organizational costs. An important disadvantage with PaaS is that it is based on virtualized technology, so we have less control over the data processing, and it is less flexible than the IaaS cloud model.

7.3 ARCHITECTURE OF CLOUD COMPUTING

The architecture of cloud computing comprises many loosely coupled cloud components and consists of three parts (Figure 7.6):

- Front end;
- Back end; and
- Virtualization.

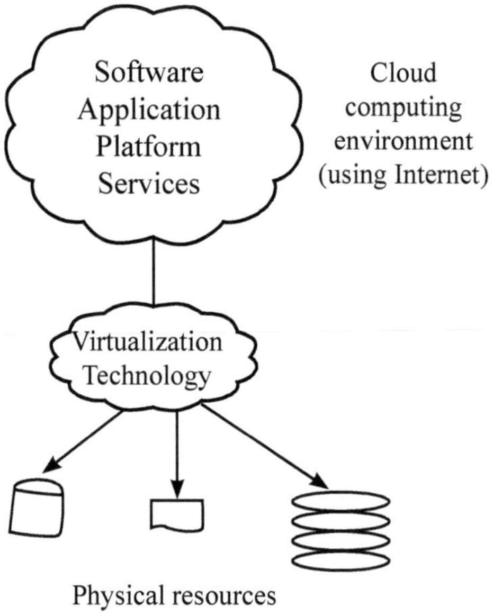

FIGURE 7.6 Cloud computing architecture.

Internet plays an important role in connecting the components of cloud computing. Below is a graphical view of cloud computing architecture:

1. **Front End:** It provides interfaces and applications necessary for accessing cloud computing platforms.
2. **Back End:** It consists of all the resources required to provide cloud computing services, such as large data storage, virtual machines, applications, services, servers, etc.
3. **Virtualization:** The important principle behind cloud computing is virtualization. The hypervisor is a low-level program that acts as a virtual machine manager, allowing the sharing of a single physical instance of cloud resources among various users.
4. **Management Software:** To maintain and configure the infrastructure in a cloud computing environment.
5. **Deployment Software:** To distribute and integrate various applications on the cloud.
6. **Network:** It allows applications and services connect over the Internet.

7.4 INTERNET OF THINGS (IOT)

The internet of things (IoT) is a system that operates without direct human-to-human or human-to-computer interaction. Instead, it relies on unique identifiers assigned to mechanical and digital machines, allowing them to interact and transfer data over a network.

The evolution of IoT is supported by the convergence of multiple technologies, real-time analytics, machine learning (ML), commodity sensors, and embedded systems. The IoT infrastructure comprises embedded systems, wireless sensor networks, control systems, automation, and other basic requirements necessary for enabling the IoT. IoT technology supports various applications, such as smart home lighting fixtures and home security systems. However, despite its numerous advantages, there are still areas of concern, such as privacy and security, which require further exploration (Figure 7.7).

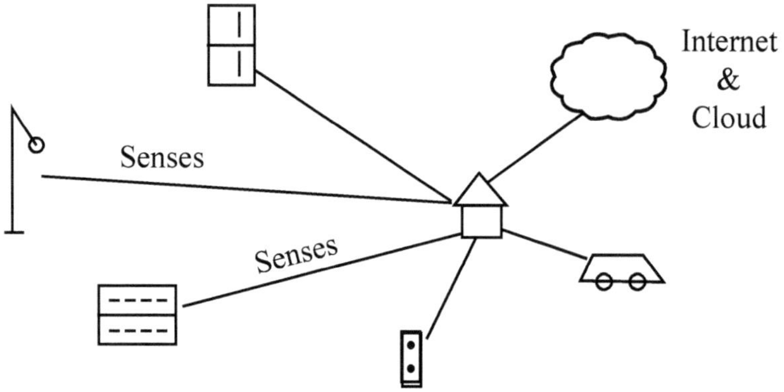

FIGURE 7.7 Internet of things.

Important pillars that support IoT infrastructure are platforms, access technologies, data storage and processing of data, analytics of data, and security. The differences between IoT and cloud computing are listed in Table 7.1.

TABLE 7.1 The Differences between IoT and Cloud Computing

Parameter	Internet of Things	Cloud Computing
Big data	It is a source for big data.	Means to manage big data over internet.
Reachability	Have limitation	Far spread, wide
Storage	Very less	Very large storage space.
Virtualization technology	Not supported	Based on the concept of virtualization.
Computing capabilities	Very limited	Unlimited
Components	Support and works on hardware components.	Support virtual machines.

7.4.1 VARIOUS IOT APPLICATIONS

1. **Smart Home:** IoT supports applications like smart homes, including the ability to switch off lights and open your home for friends and relatives when you are away. This helps simplify and convenience our lives by saving time, energy, and money for you.

Various smart home companies such as Nest, Ecobee, Ring, and August are available to turn your plans into reality.

2. **Wearables:** These devices have seen explosive demand in the market, and companies like Google and Samsung offer these devices, which are equipped with sensors and software to collect data and information about users. This data is then pre-processed to filter essential user data.

3. **Cars Connected Solution:** Automotive digital technology has also focused on optimizing the internal functions of vehicles with the help of IoT. A connected car is a type of vehicle that is designed to provide comfort to passengers by utilizing various sensors and internet connectivity.

   Major brands in this field, such as Tesla, BMW, Apple, and Google, are working towards initiating the next revolution in automobiles.

4. **Internet for Industry:** IoT is used to empower industrial engineering with sensors, software, and big data analytics, supporting the development of intelligent machines. These machines are known for their accuracy and consistency in handling data, which helps companies eliminate inefficiencies and achieve better quality control and sustainability. There are various applications available to track goods, share real-time inventory information between suppliers and retailers, and support automated delivery.

5. **Smart Cities:** It is another powerful application supported by IoT. Various smart city applications include surveillance, transportation automation, energy management systems, water distribution, pollution monitoring, traffic congestion, parking management, security, climate, and environmental monitoring, and more.

6. **Smart Technology in Agriculture:** Smart agriculture and farming is one of the fastest growing fields and requires new inventions. IoT has revolutionized farming by assisting farmers in sensing soil moisture, soil nutrients, controlling water levels for plant growth, and customizing fertilizers, among other things.

7. **Smart Retailing:** We can utilize the potential of IoT in the retail sector as well, as it offers retailers an opportunity to connect with

their customers and enhance the in-store experience. Retailers can utilize smartphones to interact with customers.

8. **Healthcare Sector:** The IoT has enormous power to connect with the healthcare system and smart medical devices, empowering people to live healthier lives. The collected data will also help in the personalized analysis of an individual's health.

7.4.2 ROLE OF CLOUD COMPUTING IN THE IOT REVOLUTION

The IoT has enormous potential to work with cloud computing in order to transform daily tasks. The IoT consists of various objects such as physical devices, vehicles, and buildings, all equipped with embedded electronic devices, software, sensors, and network connectivity. This allows objects to collect, send, and receive data.

Cloud computing is a model of computer networking that allows users to access data and programs from a centralized, organized computing resource that can scale according to demand. There are three main types of cloud deployments: public, private, and hybrid:

1. **Private Cloud Services:** It is a secure model of cloud computing that offers services exclusively for specified organizations.
2. **The Public Cloud Service:** It is a cloud network that allows for data sharing with other organizations. It also enables data transfer over a public network via the internet. Third-party providers are allowed in this network to deliver cloud services over the internet and are typically charged based on user requirements.
3. **Hybrid Cloud:** It is a cloud computing environment that combines on-premises, private cloud, and third-party public cloud services.

Any object, whether artificial or natural, that can be assigned an IP address and has the ability to transfer data over a network fall under the application of IoT. Cloud computing and the IoT both have the potential to increase efficiency in everyday tasks and are complementary to each other. The IoT generates significant amounts of data, and cloud computing provides a pathway for the data to travel. Data access is made available for cloud users, enabling better collaboration between developers who can immediately share and use data available on the cloud environment remotely. Cloud computing is important for the IoT due to the following reasons:

1. **Cloud Computing Act as a Catalyst:** The usage of IoT in the cloud computing environment has acted as a catalyst for the development and deployment of various IoT applications. The collaboration of cloud computing and IoT are two future internet technologies that provide the platform for success.
2. **Cloud Computing Space:** IoT with public cloud services is very helpful in the IoT field, as it provides easy third-party access to the infrastructure. The need for IoT devices today is to store user data, which can be provided by cloud providers like StoneFly and Microsoft Azure.
3. **Cloud Computing in Performance:** Large data produced by IoT devices needs intense performance to interact and easily connect with one another. Therefore, sharing information through the cloud provides the necessary connectivity to share information between the devices, which ultimately increases overall performance.
4. **Pay-as-per-Demand:** Cloud computing provides us with the facility to increase or decrease our requirements based on our usage, and we only have to pay for what we use. IoT provides data in the form of SMS or mobile notifications in various formats, making a cloud computing environment necessary for IoT applications.
5. **Increase Availability of Data:** Cloud technology empowers IoT for regular appliances such as air conditioners, refrigerators, etc. The cloud offers vast storage and data availability, and the use of cloud with IoT also increases.
6. **Provides Security and Privacy:** IoT is incomplete without security, but the collaboration of both Cloud has made IoT more secure with preventive, investigative, and preventive measures. It has the ability to provide users with strong security measures through authentication and encryption protocols.

7.4.3 CHALLENGES OF CLOUD AND IOT TOGETHER WORKING

1. **Data Handling:** Handling a large amount of data that evolves due to the merging of IoT with cloud is a significant challenge for managing big data.
2. **Protocols Used in Networking and Communication:** In Cloud and IoT working infrastructure, a large number of machine-to-machine communications with various types of devices, having various protocols, are used. So, managing a variety of protocols is not an easy task.
3. **Management of Sensor Networks:** Processing and management of a large amount of sensor data and network is not an easy task, as it significantly increases the burden of cloud computing in handling data. Additionally, cloud computing introduces new challenges in aggregating sensor data within a cloud environment, which can lead to security and privacy concerns.
4. **Security and Privacy Issues:** The IoT devices have allowed us to automate tasks, but there are various security and privacy concerns associated with them. While the Cloud offers robust encryption and authentication protocols, making it a secure platform for users, there are still concerns regarding identity hiding and data misuse. The availability of data on the cloud makes this a significant concern. Therefore, the merging of IoT applications with cloud platforms is a topic of research from a security perspective.

7.4.4 ROLE OF ARTIFICIAL INTELLIGENCE (AI)

AI is used to solve and perform complex and critical cognitive tasks. It makes things possible that were not even possible before the revolution of technology. With the use of inference rules like forward and backward chaining, it tries to conclude results and mimic the human brain with a structured approach. AI is applicable to process large volumes of data received from various sources such as online applications, sensors, and structured and unstructured forms of data from social media. ML, incorporated with IoT and cloud computing, is used to analyze patterns and retrieve reliable and relevant data in various decision-making situations. The cloud plays an important role in enhancing the capability of applications that incorporate AI. Artificial intelligence (AI) includes ML, speech recognition, and other technologies used by cloud providers to offer API-based applications that are intelligent and AI-enabled. IoT connects AI-enabled devices available on the cloud environment to provide better control over home and business appliances.

7.5 CONCLUSION

The internet of things (IoT) is transforming the world by offering multiple solutions to various problems. It works more efficiently when combined with cloud computing applications. IoT monitors data flow and integrates control, such as traffic-congestion avoidance techniques, enabling dynamically and efficiently managed transport networks. IoT with cloud computing applications is also involved in implementing services like IoT-based medical monitoring, utilizing cloud computing methodology in the hospital information system [1–18].

One of the major challenges in integrating cloud and IoT is the wide heterogeneity of devices, operating systems, platforms, and available services. Cloud-IoT allows for the design of novel applications where IoT serves as the main source of big data, and the cloud helps store data for long periods and enables complex analysis.

Artificial intelligence (AI) is currently dominated by information technology, as it provides better services to users. To address current and future challenges, key technologies such as cloud, IoT, and software-defined networking (SDN) are emerging as future applications. Embedding AI with these innovative technologies brings various benefits and enhances efficiency levels. This helps make systems reliable and improves the performance of data generated from heterogeneous devices, which are manipulated, stored, managed, and analyzed. The overall performance of AI with IoT and cloud technology is challenging but plays an important and prominent role in technology-oriented results.

KEYWORDS

- **artificial intelligence**
- **cloud computing**
- **healthcare sector**
- **infrastructure-as-a-service**
- **internet of things**
- **platform-as-a-service**
- **software-as-a-service**
- **software-defined networking**

REFERENCES

1. Li, J. H., Zhao, S., & Wang, Y. R., (2014). Using cloud computing and the internet of things to construct vehicle condition fault diagnosis platform. *Applied Mechanics and Materials, 574.*
2. Mao, Y., (2014). *Applied Mathematics and Algorithms for Cloud Computing and IoT.*
3. Wang, Y., & Wang, X., (2014). The novel analysis model of cloud computing based on RFID internet of things. *Journal of Chemical and Pharmaceutical Research, 6*(6).
4. Talia, D., (2014). Towards internet intelligent services based on cloud computing and multi-agents. *Advances onto the Internet of Things*. Springer International Publishing.
5. Bonomi, F., (2014). Fog computing: A platform for internet of things and analytics. *Big Data and Internet of Things: A Roadmap for Smart Environments*. Springer International Publishing.
6. Soldatos, J., (2014). Design principles for utility–driven services and cloud–based computing modelling for the internet of things. *International Journal of Web and Grid Services, 10*(2).
7. Xie, F., & Liang, C. Z., (2014). Research of internet of things based on cloud computing. *Applied Mechanics and Materials, 443.*
8. Fang, S., (2014). An integrated system for regional environmental monitoring and management based on internet of Things. *IEEE Trans. Industrial Informatics, 10*(2).
9. Ding, H. J., (2014). Traffic flow data collection and signal control system based on internet of things and cloud computing. *Advanced Materials Research, 846.*
10. Biswas, A. R., & Giaffreda, R., (2014). IoT and cloud convergence: Opportunities and challenges. *Internet of Things (WF-IoT), IEEE World Forum on IEEE.*
11. Antoine de Saint-Exupery (2009). European Research Projects on the Internet of Things (CERP-IoT) Strategic Research Agenda (SRA). *Internet of Things – Strategic Research Roadmap*. Google Scholar.

12. Armbrust, M., Fox, A., Griffith, R., Joseph, A. D., Katz, R., Konwinski, A., Lee, G., et al., (2010). A view of cloud computing. *Communications of the ACM 53*(4), 50–58. [CrossRef]. Google Scholar.
13. Ali Sunyaev. (2005). International Telecommunication Union (ITU), internet reports. *The Internet of Things.* Google Scholar.
14. Hancke, G. P., Markantonakis, K., & Mayes, K. E., (2010). Security challenges for user-oriented RFID applications within the internet of things. *Journal of Internet Technology, 11*(3), 307–313. Google Scholar.
15. Presser, M., Barnaghi, P. M., Eurich, M., & Villalonga, C., (2009). The SENSEI project: Integrating the physical world with the digital world of the network of the future. *IEEE Communications Magazine, 47*(4), 1. [CrossRef] [Google Scholar].
16. Broll, G., Rukzio, E., Paolucci, M., Wagner, M., Schmidt, A., & Hussmann, H., (2009). Pervasive service interaction with the internet of things. *IEEE Internet Computing, 13*(6), 74–81. [CrossRef] [Google Scholar].
17. Hong, S., Kim, D., Ha, M., Bae, S., Park, S. J., Jung, W., & Kim, J., (2010). SNAIL: An IP-based wireless sensor network approach to the internet of things. *IEEE Wireless Communications, 17*(6). [CrossRef] [Google Scholar].
18. Atzori, L., Iera, A., & Morabito, G., (2010). The internet of things: A survey. Computer *Networks, 54*, 2787–2805. [CrossRef] [zbMATH Google Scholar].

CHAPTER 8

BLOCKCHAIN-BASED MODEL TO PRESERVE AUTHENTICITY OF JUDICIAL EVIDENCE

SUMIT KUMAR RANA[1] and SANJEEV KUMAR RANA[2]

[1]*Assistant Professor, Panipat Institute of Engineering and Technology, Samalkha, India*

[2]*Department of Computer Science and Engineering, Maharishi Markandeshwar (Deemed to be University), Mullana, Ambala, Haryana, India*

ABSTRACT

In this era of digitization, every domain is being affected by technical advancements. The judicial system is one domain that needs improvement in its process of storing and securing evidence in an efficient way. As evidence may be accessed by multiple parties connected to a particular investigation, the originality of the evidence must be maintained. To preserve the originality of an evidence, the entire process from evidence collection to submission in court must be recorded in a tamperproof digital ledger. A combination of decentralized access control mechanism and an immutable ledger that stores the entire lifecycle of the evidence can be the solution. Blockchain technology can be very fruitful in this scenario as it has the features required to maintain the authenticity of the evidence. In this chapter, a blockchain technology supported novel method is proposed to avoid tampering and maintain the authenticity and integrity of the evidence.

The Fusion of Artificial Intelligence and Soft Computing Techniques for Cybersecurity.
M. A. Jabbar, Sanju Tiwari, Subhendu Kumar Pani, & Stephen Huang (Eds.)

8.1 INTRODUCTION

One of the main pillars of any democratic system is its judicial system. It is responsible for applying and defending the law in legal cases or disputes. If the judicial system is functioning properly, it can help resolve several issues within the system. However, there are various challenges in judicial systems, such as delays in judgment, language barriers, and warrant issuance, among others. One common problem that underlies all these issues is the preservation of evidence integrity and security. Evidence serves as the basis for fair judgment. When a case is brought to court, the evidence collected for that specific case plays a crucial role in the final verdict [1]. The process of evidence collection and management is known as the chain of custody, as depicted in Figure 8.1. It involves multiple steps, starting from evidence collection to its submission in court during the trial. This chain documents every stage of the investigation, including evidence collection, analysis, and details of who accessed it and when, in chronological order. These documents are presented during the court hearing in front of the jury, and the prosecution lawyer aims to establish the relevance of the evidence to the suspect.

FIGURE 8.1 Conventional evidence collection and management process.

On the other hand, the defense lawyer looks for loopholes in the chain of custody process so that they can prove that the evidence was not handled properly or that false evidence was created to make the suspect appear guilty [2]. In order for evidence to be legally admissible in court, it must be proven to be authentic and in its original form. Therefore, this process must have the following features to ensure its legal acceptability:

i. **Integrity:** It means protecting evidence from being altered by unauthorized entities.

ii. **Provenance:** It means recording the history of accesses to evidence, such as who and when accessed the evidence.

iii. **Authentication:** There must be a way to authenticate the various parties involved in the investigation of the case and the evidence.

iv. **Verifiable:** The process must be verifiable at any instant of time.

Judges or juries cannot conduct their own process of investigation, so they have to make a decision based on the evidence submitted in court [3]. However, if this evidence is manipulated or tampered with, their decision would challenge the integrity of the most trusted component of the democratic system.

8.1.1 PROBLEMS WITH TRADITIONAL JUDICIAL APPROACH

In the traditional approach, a form is filled out manually to document all the activities that have occurred, along with the corresponding evidence, from the time of collection until submission in court. The following information can be recorded as the evidence progresses through the investigation process:

- Date and place of origin of the evidence;
- Physical description of the evidence;
- When, how long, and who handled the evidence with all the details;
- Unique identification for each person involved in the investigation;
- Process by which evidence was transferred from one person to another person.

Few issues that can lead to a situation where important evidence might not be admissible in court and render it legally worthless are discussed below.

In the current judicial system, a plaintiff can file a complaint through an advocate. For a common man, the legal process ends up being expensive due to a lack of sufficient knowledge about legal procedures. As a result, they have to rely entirely on advocates [4]. This blind faith can lead to various consequences such as being cheated or facing inappropriate expenses.

1. **Accountability:** No court official takes responsibility for delays in court proceedings, often justifying themselves with work overload. However, in the end, it is the common man who faces the consequences [5].

2. **Provenance:** Evidence can be easily manipulated. With the current system, we cannot trace the evidence back to their original form [6].

3. **Transparency:** The current system lacks transparency in legal documents and evidence, allowing court officials to misuse information without the knowledge of other stakeholders.
4. **Data Integration:** The judicial system maintains the records of their respective zones. Therefore, integrating records becomes difficult when a case involves multiple zones.
5. **Scalability:** If the case spans over multiple states or countries, scalability will be the issue to manage the legal process. We need a technology that can resolve the issues mentioned above and support the process of evidence management and submission. Blockchain is the technology that can help in achieving this objective.

8.2 ROLE OF BLOCKCHAIN TECHNOLOGY

Blockchain is a chain of blocks that are connected with the help of hashing technique, as shown in Figure 8.2. Blockchain technology supports a distributed ledger, which is known for its immutable, tamperproof, and decentralized nature.

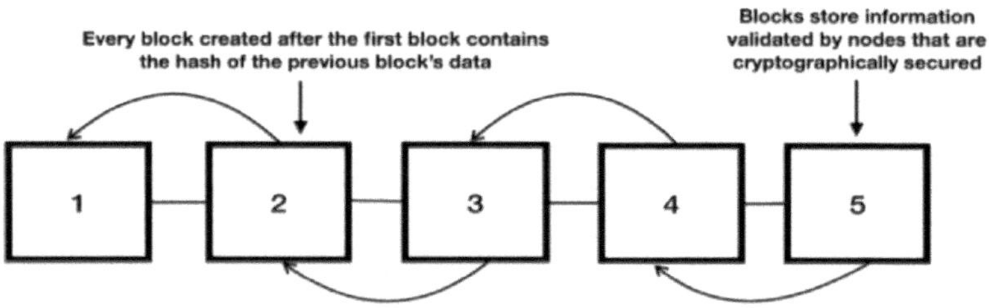

FIGURE 8.2 Conceptual illustration of blockchain.

Authors surveyed conventional schemes for sharing medical records and identified multiple issues such as trust, centralization of data, and privacy. As a result, the authors focused on a blockchain-supported approach for medical data sharing. Both public and private blockchains were analyzed, and potential research topics on blockchain-based medical data sharing were discussed [7].

The main concepts that form the foundation of any blockchain-supported application are transactions, peer-to-peer network, consensus algorithm, decentralized ledger, and smart contracts. Transactions refer to any type of interaction between nodes in a peer-to-peer network, including cryptocurrency transfers, file ownership transfers, data storage, and data access. A peer-to-peer network consists of nodes with the same capabilities or resources, without any client or server distinction. The consensus algorithm is the technique used by nodes in the peer-to-peer network to make decisions regarding the approval or denial of transactions. The decentralized ledger stores all executed transactions in the network and is shared among all nodes in the peer-to-peer network [8]. Smart contracts are computer programs designed to support agreements between communicating parties, triggered by specific events in the system.

There are various applications of blockchain technology in multiple domains like insurance, digital asset management, etc. One of the most striking examples of the technology's growth is its influence on the judicial system. Perhaps because this distributed ledger technology can store the documents in digital form and protect its integrity. This is based on peer-to-peer networks so every party can share or exchange data. All data is cryptographically secured. Every access to data is recorded on the ledger so full transparency is there. Any type of attack on integrity of data can be traced and verified. A model is proposed in which blockchain technology is integrated with back-end data storage. That data storage will be used to store the digital evidence and all evidence access transactions can be stored on the blockchain.

8.3 RELATED STUDY

Authors have explored the importance of video evidence in investigations. However, tampering with video evidence poses a serious problem. To address this, the authors have proposed a model supported by blockchain for integrity verification. In this model, a hash-based method is used to generate a video integrity code. If any of the video segments are tampered with, this integrity code will change, allowing for easy detection of tampering by comparing the two video integrity codes. Comparative analysis has shown that the proposed model performs better in terms of video evidence security [9].

Furthermore, the authors have proposed a model for accumulating evidence for investigations from a cloud environment. This model utilizes SDN and distributed ledger technology to preserve the evidence.

Additionally, a new algorithm has been introduced to provide security to the collected data. The complete system has been simulated using Java language and NS3. Evaluation analysis of the proposed model has demonstrated its efficiency compared to centralized systems [10].

Researchers have developed an approach for IoT forensics that maintains the privacy of data shared by individuals. The PRoFIT methodology was employed in conjunction with the discussed approach. A model for accumulating digital evidence from IoT environments was studied. This approach has been evaluated by applying it in different scenarios with varying privacy needs. The evaluation has shown that the proposed model is able to maintain a balance between secrecy and the principles of internet of things (IoT)-based investigation [11].

The authors addressed the problem of discrimination and justice faced by customers in the insurance domain. They developed a scheme to provide security and immutability to stored insurance records and documents. Blockchain technology was employed in conjunction with the interplanetary file system (IPFS) for this scheme. This combination provides security and immutability while also avoiding the problem of repudiation. In the future, researchers are also interested in using smart contracts in the implementation of the scheme [12].

Researchers investigated the variety of forensic evidence, information formats, and other complexities in the IoT environment. Their main interest was to accumulate information and objects from various devices connected to the IoT network. After collecting the objects and information, researchers examine the relationship between the evidence and then record it on the blockchain-based forensic model [13].

Authors studied the problems faced with a centralized network when used for storing or exchanging the medical information of a patient. They found various issues such as confidentiality, unauthorized access, and lack of historical data. The authors proposed a model that combines blockchain technology with cryptography to address the aforementioned issues of a centralized network. All information is uploaded with the agreement of the concerned parties, and data can be accessed based on the consensus of the concerned parties [14].

Researchers explored the requirements of different stakeholders involved in the digital proof collection process. They also analyzed that the system must have a way to ensure the admissibility of the evidence in court. Therefore, they developed a system supported by blockchain and smart contracts that can collect, store, and share evidence with the stakeholders. This chapter also discusses the practicality of the proposed system [15].

Authors have explored the uses of blockchain technology in the judicial domain. Properties of blockchain, such as provenance, transparency, and decentralization, can help maintain the integrity and trust in the evidence collection and sharing process. The authors have proposed a blockchain-based system for electronic evidence generation. With the help of this system, judicial entities can verify the validity of any evidence presented in court trials. Analysis of the system has proven that the proposed system provides provenance, trust, and efficiency to the application area [16].

Researchers have demonstrated that by hashing data alone, we cannot protect our data because it does not provide any information regarding the time of hash computation. Therefore, an approach is needed to timestamp the hashing process. Researchers have come up with a solution based on blockchain. In the proposed approach, public blockchains were used to prevent data alteration and increase data transparency. Different stakeholders of the judicial process can track and analyze the evidence at any given moment. This model can serve as the foundation for new researchers in the same domain [17].

Authors explored the use of blockchain technology for the medical research community. They proposed a model for storing and querying pharmacogenomics data, which was demonstrated using the Ethereum blockchain and solidity-based smart contracts. Analysis of the algorithm showed that the proposed model was fast and reduced query time with a query pool of 10,000 queries. The algorithm was developed with the constraints of solidity in mind, such as the number of variables and gas requirements. The algorithm used in this model was successful in the medical domain. As future work, the authors are hopeful about implementing the same model in other domains as well [18].

Authors demonstrated the possible safety and privacy-related issues in the maintenance of judicial evidence in digital forensics. They proposed a blockchain and smart contract-based model called LEChain. This model incorporated the complete process of evidence collection and reporting. The model used an ABE scheme to ensure fine-grained access control of evidence. The proposed model also safeguards the privacy of eyewitnesses during investigation. As evidence data is uploaded on the blockchain, it cannot be modified by malicious entities. This model was implemented using the Ethereum test network [19].

8.4 PROPOSED MODEL

In the proposed model, distributed ledger technology is employed to provide security to the evidence by using the Ethereum blockchain and IPFS. Let's discuss the various concepts of the model and the workflow of the proposed model.

8.4.1 SYSTEM MODEL

The key components of the projected model are judicial evidence, the public blockchain platform Ethereum, the IPFS for data storage at the back end, and various users of the system. The aforementioned components are connected in such a way that information can be exchanged among them as per requirement. Let's discuss these components:

1. **Judicial Evidence:** It is the document used as proof for fair judgment in any court case. It is the most significant component of the proposed model. Various entities will access evidence during the processing of a court case. The main objective is to ensure the security and integrity of these evidence, as tampered or manipulated evidence can lead to incorrect judgments [20]. Such wrong decisions can undermine trust in the judicial system. Therefore, a decentralized and transparent system is built using the Ethereum blockchain platform.
2. **Ethereum Blockchain:** Decentralization of powers is required to build a trusted and transparent system. This decentralization can be achieved with the help of a public blockchain platform named Ethereum. Ethereum has a distributed ledger that is shared among all the stakeholders of the peer-to-peer network. All transactions, such as upload, delete, or access, are stored in this ledger. The ledger is immutable, which means that once something is written on it, it cannot be changed or deleted [21]. However, due to the small block size of the blockchain, evidence cannot be stored on it. Therefore, the IPFS is used to satisfy our off-chain storage needs.
3. **IPFS:** As our objective is to build a decentralized system, we also need a storage mechanism that is not centralized. The IPFS can be employed to fulfill this need. IPFS stores data in a distributed

manner by utilizing nodes located at different geographical locations. When data is stored on this storage system, a unique hash is returned, known as the content address (CID) of the document. This CID can be used to retrieve the file in the near future. The same CID is returned if files with the same data are uploaded, thus avoiding data duplication. In the proposed model, judicial evidence will be stored on IPFS [22].

4. **System User:** Users of this proposed system will include officials from the police department, prosecution lawyers, defense lawyers, and others. Each user will be provided with a pair of keys that can be used to access the evidence related to a specific case. These keys will authenticate the user and prevent repudiation.

8.4.2 WORKFLOW

The workflow of the model can be divided into two phases based on the communication. Phase I describes the communication between the evidence uploader (UE) and the system, as shown in Figure 8.3. Phase II describes the communication between the access requester (RA) and the system, as shown in Figure 8.4. Before these phases, some assumptions are made: every user is a registered user, and the system cannot be accessed without registration.

- **Phase I:** U_E is an official from the police department who will upload the evidence document (D_E) on the IPFS. First of all, U_E will attach a digital signature ($Sign_D$) to D_E and submit this combination to the system for uploading on IPFS. Before uploading on the IPFS, the system will recalculate $Sign_D$ and compare it with the received $Sign_D$ to verify that the document is intact and to avoid repudiation. If both $Sign_D$ values are the same, D_E will be uploaded in encrypted form on IPFS. You will receive a hash value known as the content address (A_C). This A_C is used to retrieve the actual content of the evidence stored on IPFS. The metadata of this uploading transaction (T_{iM}) is stored on the blockchain. U_E will receive a block number (B_i) where the metadata of transaction T_i is stored on the blockchain.

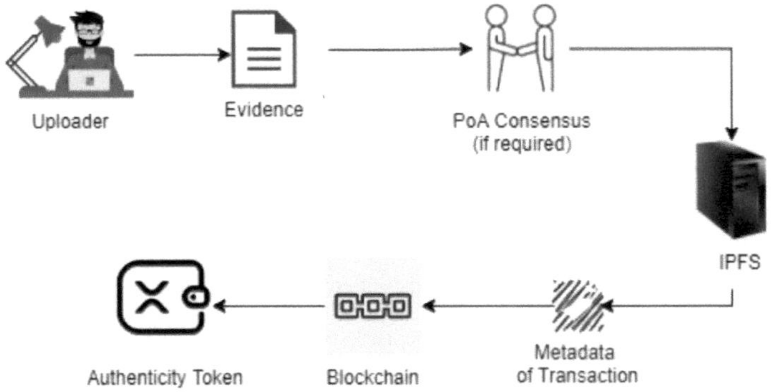

FIGURE 8.3 Workflow Phase I.

- **Phase II:** This phase begins when any registered user, R_i, wants to access the evidence, D_E. R_i will log in to the system using the provided credentials. Then, $_{Ri}$ will be able to see the list of evidence related to his case. If R_i clicks on any evidence to view it, a request will be sent to the other concerned parties of the respective case. These concerned parties will approve or deny the request using proof of authority (PoA). After approval, R_i will receive an access code (A_C) for that evidence. By using this A_C, R_i can access that particular D_E. The metadata of this access transaction is stored on the blockchain.

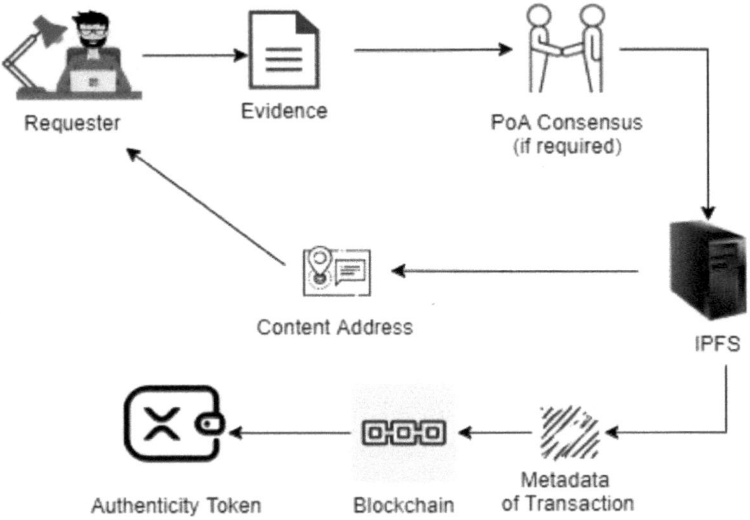

FIGURE 8.4 Workflow Phase II.

8.5 ADVANTAGES OF PROPOSED MODEL

The proposed model, which employs blockchain technology in the legal process, offers various advantages that are discussed below:

1. **Evidence Sharing Without Trusted Third Party:** With the proposed model, all evidence can be easily shared with all entities involved in the case without the need for any third party to validate transactions. Therefore, any trust issues or failures can be avoided [23].
2. **Retaining Evidence Integrity:** Blockchain technology has a feature of immutability. Because of this feature, if any transaction is written on the ledger, it can't be deleted or altered. So, if the information related to evidence is once stored on this, nobody can manipulate the information. This is how the integrity of the evidence can be retained [24]. This can be very important in the current digital world where the reliability of digital media is at stake due to fake photos or videos in circulation.
3. **Digital Evidence Storage:** Maintenance of physical documents over time is very difficult. For example, paper documents go through wear and tear with time, and images fade away. Therefore, it is better to store these documents in digital form. This can be easily done with the help of the proposed model [25, 26].
4. **Evidence Tracking:** All evidence has been uploaded to the IPFS, with its metadata stored on the Ethereum blockchain. The Ethereum blockchain will track all access transactions, including information such as who accessed it and when, from its provenance time to the current instance of time [27].
5. **Reduction of Fraud by Increasing Transparency:** All access to any evidence will only be allowed with the consensus of the involved parties, and access transactions will be updated in the ledgers of all parties. Therefore, this level of transparency will reduce the chances of fraud or manipulation with the evidence [28].
6. **Multi-Country Investigation:** There might be some cases where multiple countries need to cooperate to carry out the investigation of any case. In those cases, sharing and maintaining evidence could create problems. With the proposed model, evidence sharing can be easily performed in cross-border investigations [29].

8.6 CONCLUSION

A democratic system is dependent on its judiciary system. If the judiciary system fails, it is a failure of the democratic system. The current judicial system is being exploited by advocates and other officials. It suffers from various issues that can result in injustice or delays in justice. It lacks transparency and accountability. Blockchain technology offers various features that can help resolve the issues of the conventional judiciary system. Blockchain utilizes a distributed ledger that is shared among all stakeholders, ensuring complete transparency. Cryptographic techniques are employed to ensure the security and privacy of information. Immutability is a crucial feature that makes it suitable for the judiciary domain, as any transaction recorded in the ledger cannot be changed, deleted, or repudiated. Integrating blockchain technology with the judiciary system will enhance security and traceability. This will significantly improve trust in the judiciary system, as no one will be able to misuse the system for personal gain. The ultimate objective of regaining the trust of the common man in the judicial system can be achieved with the help of the proposed blockchain solution.

KEYWORDS

- **blockchain**
- **chain of custody**
- **cryptographic techniques**
- **decentralized access control**
- **interplanetary file system**
- **judicial evidence**
- **multi-country investigation**
- **traditional judicial approach**

REFERENCES

1. Nieto, A., Roman, R., & Lopez, J., (2016). Digital witness: Safeguarding digital evidence by using secure architectures in personal devices. *IEEE Netw., 30*(6), 34–41.
2. Cebe, M., Erdin, E., Akkaya, K., Aksu, H., & Uluagac, S., (2018). *Block4forensic: An Integrated Lightweight Blockchain Framework for Forensics Applications of Connected Vehicles* (pp. 50–57). arXiv.
3. Tian, Z., Li, M., Qiu, M., Sun, Y., & Su, S., (2019). Block-DEF: A secure digital evidence framework using blockchain. *Inf. Sci. (Ny)., 491*, 151–165.
4. Lin, X., Chen, T., Zhu, T., Yang, K., & Wei, F., (2018). Automated forensic analysis of mobile applications on Android devices. *Proc. Digit. Forensic Res. Conf. DFRWS 2018 USA, 26*, S59–S66.
5. Shafarenko, A., (2020). *A PLS Blockchain for IoT Applications: Protocols and Architecture*. arXiv.
6. Guo, R., Shi, H., Zhao, Q., & Zheng, D., (2018). Secure attribute-based signature scheme with multiple authorities for blockchain in electronic health records systems. *IEEE Access, 6*, 11676–11686.
7. Jin, H., Luo, Y., Li, P., & Mathew, J., (2019). A review of secure and privacy-preserving medical data sharing. *IEEE Access, 7*, 61656–61669.
8. Pourvahab, M., & Ekbatanifard, G., (2019). An efficient forensics architecture in software-defined networking-IoT using blockchain technology. *IEEE Access, 7*, 99573–99588.
9. Ghimire, S., Choi, J. Y., & Lee, B., (2020). Using blockchain for improved video integrity verification. *IEEE Trans. Multimed., 22*(1), 108–121.
10. Pourvahab, M., & Ekbatanifard, G., (2019). Digital forensics architecture for evidence collection and provenance preservation in IAAS cloud environment using SDN and blockchain technology. *IEEE Access, 7*, 153349–153364.
11. Nieto, A., Rios, R., & Lopez, J., (2018). IoT-forensics meets privacy: Towards cooperative digital investigations. *Sensors (Switzerland), 18*(2).
12. Sun, J., Yao, X., Wang, S., & Wu, Y., (2020). Non-repudiation storage and access control scheme of insurance data based on blockchain in IPFS. *IEEE Access, 8*, 155145–155155.
13. Li, S., Qin, T., & Min, G., (2019). Blockchain-based digital forensics investigation framework in the internet of things and social systems. *IEEE Trans. Comput. Soc. Syst., 6*(6), 1433–1441.
14. Lusetti, M., Salsi, L., & Dallatana, A., (2020). A blockchain-based solution for the custody of digital files in forensic medicine. *Forensic Sci. Int. Digit. Investig., 35*, 301017.
15. Aparecido, P. B. C., Gonçalves, R. F., Sérgio De, A. I. P., Reis, J. Z., & Dolce, U. G. J. M., (2020). Smart contracts applied to a functional architecture for storage and maintenance of digital chain of custody using blockchain. *Forensic Sci. Int. Digit. Investig., 34*.
16. Chen, S., Zhao, C., Huang, L., Yuan, J., & Liu, M., (2020). Study and implementation on the application of blockchain in electronic evidence generation. *Forensic Sci. Int. Digit. Investig., 35*, 301001.

17. Burri, X., Casey, E., Bollé, T., & Jaquet-Chiffelle, D. O., (2020). Chronological independently verifiable electronic chain of custody ledger using blockchain technology. *Forensic Sci. Int. Digit. Investig., 33*(xxxx).
18. Gürsoy, G., Brannon, C. M., & Gerstein, M., (2019). *Using Ethereum Blockchain to Store and Query Pharmacogenomics Data Via Smart Contracts* (pp. 1–11). bioRxiv.
19. Li, M., Lal, C., Conti, M., & Hu, D., (2021). LEChain: A blockchain-based lawful evidence management scheme for digital forensics. *Futur. Gener. Comput. Syst., 115*, 406–420.
20. Kosba, A., Miler, A., Shi, E., & Wen, Z., (2016). Hawk: The blockchain model of cryptography and privacy preserving smart contracts. *IEEE Symposium on Security and Privacy*, 839–858.
21. buterin, V., (2013). *A Next Generation Smart Contract & Decentralized Application Platform- Ethereum White Paper* (pp. 1–36). Ethereum.org.
22. Norvill, R., Pontiveros, B. B., State, R., & Cullen, A., (2018). IPFS for reduction of chain size in Ethereum. *IEEE Conference on IoT, Green Computing, Smart Data, Blockchain, Congress on Cybernetics* (pp. 1121–1128).
23. Simmross, U., (2014). Appraisal of scientific evidence in criminal justice systems: On winds of change and coexisting formats. *Law, Probability and Risk, 13*, 105–115.
24. Pereira, R. S., (2013). Evidence models and proof of causation. *Law, Probability and Risk, 12*, 229–257.
25. Rana, A. K., & Sharma, S., (2021). Industry 4.0 manufacturing based on IoT, cloud computing, and big data: Manufacturing purpose scenario. In: *Advances in Communication and Computational Technology* (pp. 1109–1119). Springer, Singapore.
26. Rana, A. K., & Sharma, S., (2021). Contiki cooja security solution (CCSS) with IPv6 routing protocol for low-power and lossy networks (RPL) in internet of things applications. In: Mobile Radio Communications and 5G Networks (pp. 251–259). Springer, Singapore.
27. Rana, S. K., & Rana, S. K., (2020). Blockchain-based business model for digital assets management in trust less collaborative environment. *Journal of Critical Reviews, 7*(19), 738–750.
28. Khan, S., Loukil, F., Guegan, C. G., Benkhelifa, E., & Bani, H. A., (2021). Blockchain smart contracts: Applications, challenges, and future trends. *Peer-to-Peer Networking and Applications.* Springer.
29. Dias, D., & Benet, J., (2016). Distributed web applications with IPFS. In: *ICWE* (pp. 616–619). Springer LNCS.

PART III

MACHINE LEARNING AND DEEP LEARNING FOR AGRICULTURE

CHAPTER 9

APPLICATIONS OF MACHINE LEARNING AND DEEP LEARNING TECHNIQUES IN AGRICULTURE FOR WEEDS AND CROP DISEASES DETECTION AND CLASSIFICATION

K. UPENDAR, K. N. AGRAWAL, and N. S. CHANDEL

ICAR–Central Institute of Agricultural Engineering, Nabi Bagh, Berasia, Bhopal, India

ABSTRACT

Traditional machine learning (ML) models and convolutional neural networks (CNN)-based deep learning (DL) models have been extensively used in the agriculture sector for plant disease detection, weed and plant classification. Color, texture, shape, and spectral features help to identify and classify objects into different classes. Traditional ML and DL algorithms for classification are simpler and faster than traditional scanning techniques or manual classification techniques. Feature extraction using transfer learning models (VGGNet, AlexNet, GoogleNet, and DenseNet) has attracted more attention than shape and texture feature extraction using traditional image processing techniques. Python offers several free libraries for image processing and building ML algorithms. TensorFlow is the most commonly used Python library for developing CNN-based DL models. These DL models have been converted to TensorFlow Lite and deployed on edge computing devices, such as Raspberry Pi. These

The Fusion of Artificial Intelligence and Soft Computing Techniques for Cybersecurity.
M. A. Jabbar, Sanju Tiwari, Subhendu Kumar Pani, & Stephen Huang (Eds.)

advancements in science and technology will empower robots to detect pests and apply insecticides to diseased plants in the future. Moreover, robots will apply herbicides in a site-specific manner and minimize the need for herbicides.

9.1 INTRODUCTION

Over the last two decades, advancements in opto-electric systems, coupled with improved computational capabilities, have enabled users to capture and analyze images in real time. Image processing, machine learning (ML), and deep learning (DL)-based computer vision have been continuously used for fast and accurate detection of plant diseases and weeds in a non-destructive manner within crop fields [1]. ML allows machines to learn from datasets, rather than relying on a set of rules and explicit programming. Real-time, site-specific application of pesticides or herbicides has become a highly promising technology in modern-day agriculture, thanks to advancements in sensing devices and high computational power of hardware components. Consequently, an automated detection system is crucial for identifying weeds and diseased plants for early diagnosis and immediate implementation of necessary control measures. An automated detection system saves time and assists farmers in improving crop quality and reducing disease occurrence through early diagnosis and prompt treatment [2].

The process of detecting plant diseases and weeds begins with image acquisition, followed by pre-processing and image segmentation. Various ML and DL models have been utilized to classify features extracted from segmented images into clusters [3]. Google Colab, Microsoft Azure, and Kaggle Notebooks are public cloud computing platforms that provide graphical processing units (GPU) and tensor processing units (TPU) for ML and CNN-based DL applications. Scikit-learn is an open-source Python library that assists in building regression, clustering, and classification models. TensorFlow is the most commonly used Python library for developing CNN-based DL models. ML and DL models have been applied in various agricultural applications, including crop disease identification, weed and pest classification [4].

9.2 TYPES OF FEATURES

Color, shape, texture, and spectral features of the soil background, weeds, main crops, healthy and unhealthy crops are essential for detection and classification using machine and DL models. The accuracy of color-based classification mainly depends on the color difference and the specific plant being studied. The color feature is robust to scale, position, and size [5]. The color feature is commonly used for vegetation segmentation from the soil background. The performance of several vegetation indices, such as the excess green index (ExG), excess red index (ExR), and excess green minus red index (ExGR), on vegetation segmentation accuracy has been extensively studied [6]. However, the color feature alone is not sufficient to classify the main crop from weeds or different types of weeds, as both the main crop and weeds are typically green in most cases.

Shape feature is the most commonly used for weeding or crop detection and crop species identification. A list of shape features is provided in Table 9.1. Shape features have been successfully employed for plant species recognition tasks [7, 8]. Plant leaves are typically flat, with varying vein textures and leaf surface roughness. In texture-based weed or plant disease detection, the most commonly used techniques for feature extraction are given in Table 9.1. Texture information obtained from the main crop, weeds, or diseased plants plays a significant role in classification tasks [9].

Along with color, shape, and texture features, spectral features also play a significant role in plant or weed classification problems. When the spectral reflectance of the weed class is significantly different from the main crop's spectral feature, it can be employed to distinguish weeds from the main crop [10]. Compared to shape and texture features, the spectral feature is robust to partial occlusion of leaves [11]. Researchers have investigated the use of fluorescence [12], multispectral or hyper-spectral imaging (HIS) [13], visible and near-infrared spectra [14] for the classification of plants and weeds.

9.3 TYPES OF MACHINE LEARNING (ML) MODELS

Machine learning (ML) models are divided into regression, classification, and clustering. Regression and classification techniques are called supervised learning algorithms. The class label or target is a continuous variable for regression problems (Figure 9.1). However, the class label is a categorical variable in the case of classification and clustering problems. The clustering

TABLE 9.1 Types of Features

Category	Technique
	Color Feature [15]
Color models	Red (R), green (G), blue (B) hue (H), saturation (S), value (V) luminance, chrominance, chroma (YUV)
	Shape Features [5]
Shape parameters	Perimeter, compactness, area, diameter, minor axis length, rectangularity, major axis length, circularity, eccentricity, convexity, and solidity.
Region-based descriptors	Two dimensional Fourier descriptors (FD) and Hu moment invariants.
Contour-based descriptors	Spatial position descriptor, one dimensional FD, curvature scale descriptors.
	Texture Features [5]
Statistical method	Gray level co-occurrence matrix (GLCM): 10 statistics, gray level gradient co-occurrence matrix (GGCM): 15 statistics.
Structural method	Local binary pattern (LBP)
Model-based method	Fractal dimension
Transform-based method	Gabor-based

algorithms are called unsupervised learning algorithms. Supervised ML models require features and corresponding true labels to train and test the models. Linear regression, decision trees (DT), logistic regression (LR), artificial neural network (ANN), random forest (RF), support vector machine (SVM), and K-nearest neighbors (KNN) are supervised ML models.

In the case of an unsupervised ML algorithm, class labels are not defined; instead, a number of clusters in the dataset are defined [1]. Unsupervised ML models include Density-based spatial clustering of applications with noise (DBSCAN) and K-means clustering. Several stages are involved in developing ML models. A list of supervised and unsupervised ML models is given in Figure 9.2. The different stages involved in the development of ML or DL models are given in Figure 9.3.

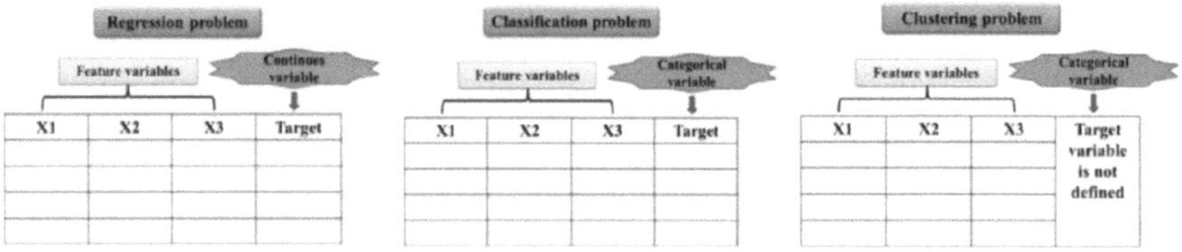

FIGURE 9.1 Types **of machine learning (ML) models.**

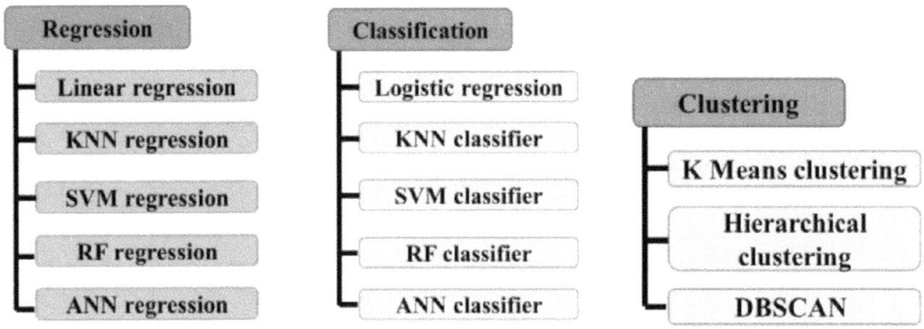

FIGURE 9.2 List of supervised and unsupervised machine learning models.

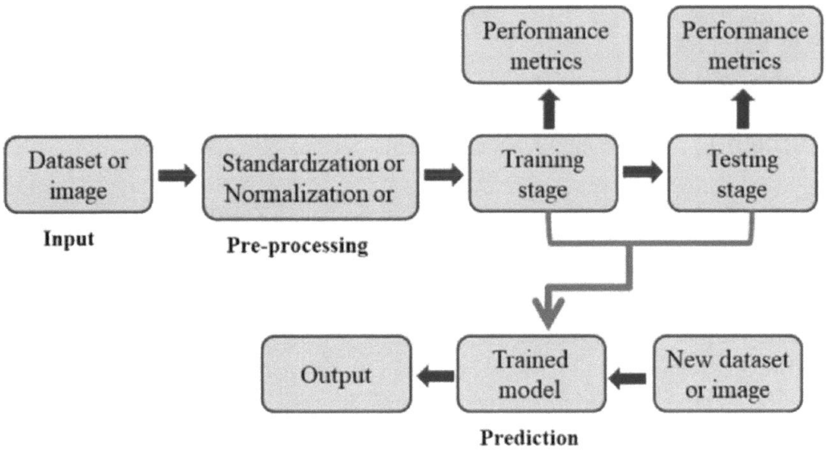

FIGURE 9.3 Stages involved in the development of machine learning or deep learning models.

9.3.1 DIFFERENCE BETWEEN TRADITIONAL MACHINE LEARNING (ML) AND DEEP LEARNING (DL)

Digital color cameras store color information in the red, green, and blue (RGB) color format. Each layer represents a gray scale image. Depending on the requirements, the RGB color model can be converted to HSV (hue, saturation, value), CIE Lab, and YUV. Meng et al. [15] reported that the Hue channel from the HSV color model performs better than the RGB and YUV color models in image processing applications under field conditions. In the case of traditional ML models, features need to be manually extracted using image processing techniques (**Figure 9.**4) and then inputted into classification, regression, and clustering models. Features are extracted from the color models using the filters mentioned in Figure 9.4 and then used for training and testing in the learning models.

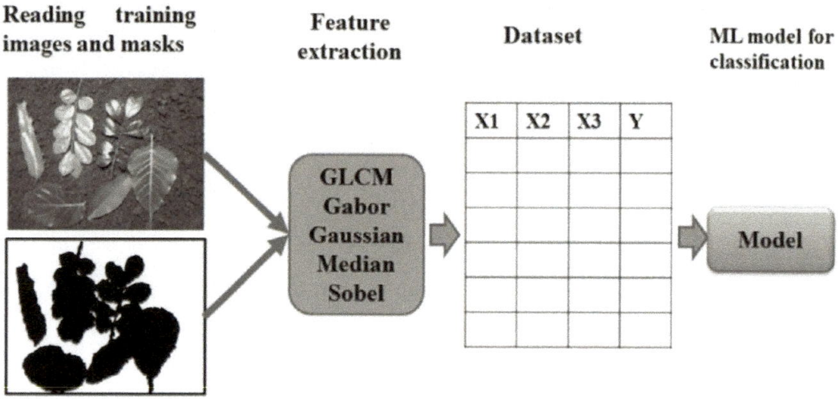

FIGURE 9.4 Stages involved in traditional machine learning approaches.

In the case of a convolutional neural network (CNN)-based DL network, different types of filters/kernels are part of the network (**Figure 9.**5), and they extract salient features from the image. These features are used as input and passed through a deep neural network for classification and regression tasks. The CNN-based DL network consists of two stages. The first stage is featuring learning, and the second stage is the classification or regression problem.

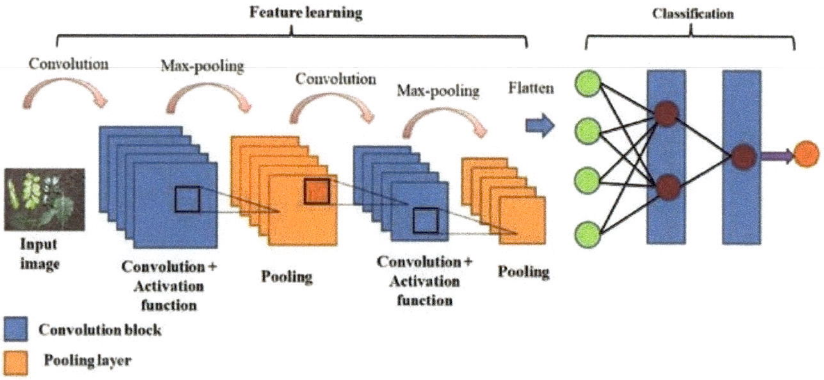

FIGURE 9.5 Stages involved in CNN based deep learning approaches.

The first stage consists of several operations such as convolution, pooling, stride, and flattening operations. In the context of CNNs, a kernel is also referred to as a filter and feature detector. Assuming square kernels, the dimension of a filter is 3×3. The filter will be convolved with the channels of the input image, as shown in Figure 9.6 (input dimensions: Height x Width x Channels). The workflow of the convolution operation is depicted in Figure 9.6.

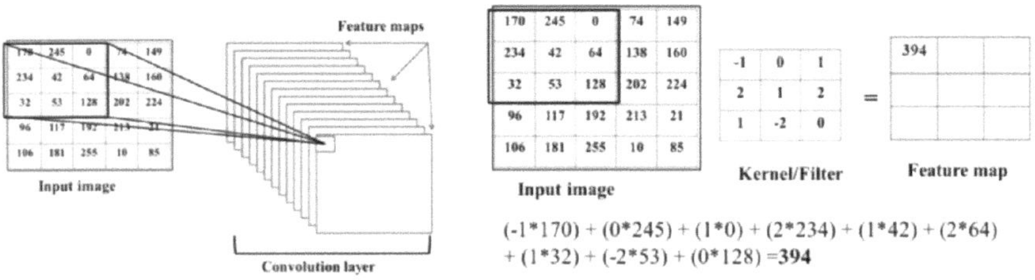

FIGURE 9.6 Working flow of Kernal or filter operation.

Pooling operations can be performed in three ways: max pooling, mean pooling, and sum pooling. In Figure 9.7, a feature map of size 5 x 5 is shown. To perform the pooling operation, an array size of 2 x 2 was selected (Figure 9.7). A 2 × 2 array size is commonly used. When the 2 x 2 array is applied to the feature map, the elements within the array are 0, 1, 0, 1. In the case of max pooling, the highest number among the four elements is selected. In this example, 1 is the highest among the remaining three elements. In the case of mean pooling, the mean of all four elements is considered. In the case of sum pooling, the sum of all elements is considered. The output of the CNN architecture is in array format. The array needs to be flattened before being sent through the dense neural networks, as shown in Figure 9.8.

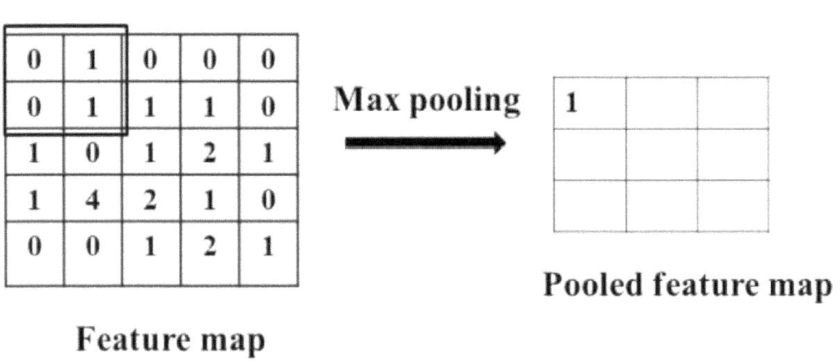

FIGURE 9.7 Working flow of max pooling operation.

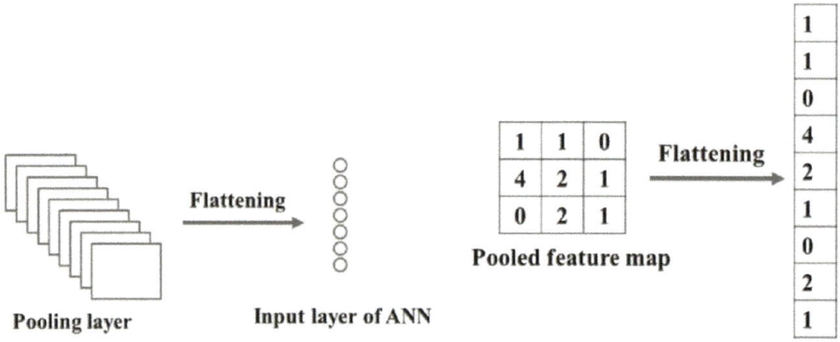

FIGURE 9.8 Working flow of flattening operation.

DL CNNs such as VGGNet, Alex Net, Google Net, and Detect Net were explored by Yu et al. [16]. These models were trained on a large set of datasets. Many open-source image datasets are available for CNN-based DL model training and testing. These datasets are manually annotated and consist of different categories of images. COCO [17], open images [18], image net [19], and Pascal VOC [20] datasets are available free of cost to the scientific community. Several researchers used these pre-trained models (**Figure 9.9**) for feature extraction and classification.

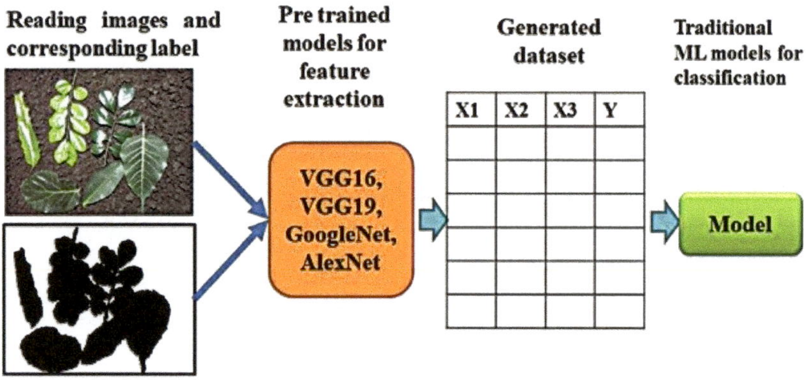

FIGURE 9.9 Pre-trained models for feature extraction and traditional machine learning models for classification.

9.3.2 *PERFORMANCE METRICS (PM) OF CLASSIFICATION ALGORITHMS*

Several performance metrics (PM) have been employed to measure the ability of ML models in regression and classification problems. There is a need to measure the performance accuracy of models with different hyper-parameters. True positive (TP), false positive (FP), true negative (TN), and false negative (FN) are part of the confusion matrix. Accuracy, precision, and recall are derived from TP, TN, FN, and FP [21]. F1-score, accuracy, precision, and recall are used as PM for classification models (**Figure 9.**10). If the model predicts weed as weed, it is called TP, whereas if weed is classified as the main crop, it is called FN. Similarly, if the main crop is classified as the main crop by the model, it is called TN, whereas if the model predicts the main crop as weed, it is called FP [22].

$$\textbf{Accuracy} = \frac{TP+TN}{TP+TN+FP+FN}$$

$$\textbf{Precision} = \frac{TP}{TP+FP}$$

$$\textbf{Recall} = \frac{TP}{TP+FN}$$

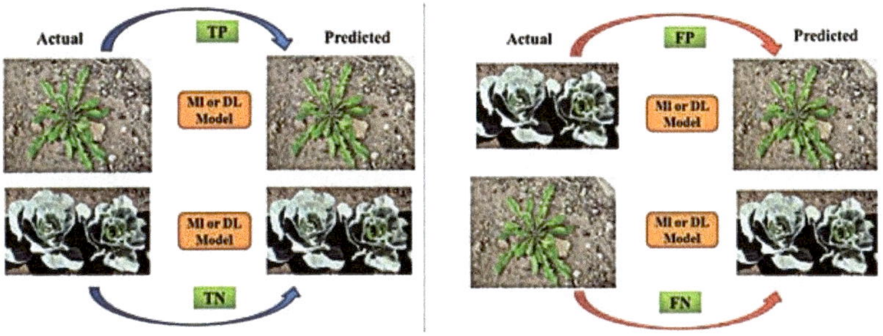

FIGURE 9.10 Performance metrics of classification problem.

9.4 WEED DETECTION

Weeds are the greatest serious hazard to crop yield, aside from diseases. The most difficult aspect of weed control is detecting and distinguishing them from crops. Herbicides are the most commonly used tool for weed control, which poses concerns. Site-specific herbicide application technology eliminates the blanket application of herbicides by spraying only on weed-infested areas. Site-specific herbicide application requires sensing and spraying units; therefore, machine vision is vital in agricultural automation.

9.4.1 TRADITIONAL MACHINE LEARNING (ML) MODELS

Co-occurrence matrix-based texture features were used to extract meaningful information from images, which were then examined to detect wild blueberry plants, bare patches, and weeds [23]. Texture features were extracted from the photos using MATLAB software and analyzed using

SAS software. Around 44 textural features were derived from the hue, saturation, intensity, and NTSC luminance (L) channels. The choice of algorithm will depend on whether the end-user application prioritizes processing speed or accuracy. The hue and saturation features of the HSI color model had a greater influence on accuracy compared to other color spaces. Rehman et al. [24] utilized color co-occurrence matrices (CCMs) to extract textural features from goldenrod in wild blueberry fields. It was reported that using hue and saturation color channels resulted in an overall accuracy of > 90% in both the training and test sets.

For weed detection in grain fields, both color and shape features were considered [25]. A normalized difference index (NDI) was created using the green and red channels of color images to distinguish between vegetation and background. The shape feature was utilized to differentiate between the main crop and weeds. The performance of the weed detection model was compared to visual surveying, resulting in an improvement in correlation from 75% to 85% when shape analysis was employed. The weed detection accuracy using Bayes rule was 74.5%, while KNN achieved a weed detection accuracy of 79.2%. By considering both the shape feature and color in crop and weed identification, the detection accuracy increased from 75% to 85%. The shape feature shows potential in enhancing performance and resilience against environmental conditions in field-deployable systems. However, occlusion remains a challenge. Further research is needed to address the computational cost of leaf segmentation and explore the combined use of shape features with color indices for maize detection.

The overlapping of leaves is another disadvantage of the shape-based weed identification approach [26]. Four types of weeds were classified based on their shape and texture [27]. A SVM was used as a classifier. First, the image background (soil) was segmented from the weed images. Then, shape and texture features were extracted from the segmented images. The model achieved a classification accuracy of 93.3%, with a processing time of 1.18 seconds for the classification algorithm. Fractal dimension, texture, and plant leaf shape features were considered for weed recognition [28]. These three features improved the classification ability of the SVM classifier, especially for small-sized datasets. Bakhshipour et al. [29] used an ANN for weed segmentation in sugar beet. They examined the potential of wavelet texture features in separating weed images from sugar beet crops. A total of 52 texture features were extracted for weed segmentation tasks, which were then reduced to 14 principal components using principal component analysis (PCA). Even in cases of substantial occlusion and overlapping of leaves, the wavelet texture features were able to efficiently differentiate weeds from the crops.

The performance of the SVM and ANN for weed detection in sugar beet fields was compared by Bakhshipour et al. [30]. Two shape features, moment invariants and Fourier descriptors (FD), were used as inputs to the SVM and ANN classification models. The SVM showed the highest classification accuracy (95.00%) compared to the ANN (92.92%). The ANN correctly classified around 93.33% of sugar beets and 92.50% of weeds, while the SVM correctly classified 93.33% of weeds and 96.67% of sugar beets.

A SVM was employed for the classification of three crops [31]. Radish, *Brassica napus* (canola), and *Zea mays* (maize/corn) crop texture features were extracted using Local Binary Pattern (LBP) operators. The dataset size was 30,000 images. The subclasses of the dataset were soil background, maize, canola, and radish. During the training process, 24,000 images were used to train the model, and 6,000 images were used for model validation. The classification accuracy of the SVM model in classifying images into four groups was reported as 91.85%. Gabor, histogram of oriented gradient feature (HOG), GLCM, hu invariant moment feature, rotation invariant LBP feature, and GGCM features were used for corn seedling and weed detection using a SVM classifier [32]. The dataset consists of 2,000 images of corn seedlings and weeds. Image background was removed by K-means clustering and the LAB color model. PCA was used for dimensionality reduction.

9.4.2 CNN-BASED DEEP LEARNING (DL) MODELS

Deep learning (DL) CNNs such as VGGNet, AlexNet, GoogleNet, and DetectNet were explored by Yu et al. [16] to identify ground ivy, dandelion, and spotted spurge in perennial ryegrass. The training method of DCNN, particularly VGGNet and DetectNet, shows potential for developing a computer vision-based site-specific herbicide controller in perennial ryegrass. Chechliński et al. [33] also investigated DL CNN models for weed identification using Raspberry Pi 3B+. The CNN architecture includes DenseNet, MobileNet, U-Net, and ResNet transfer learning models to classify four plant species at different growth stages and light conditions. The processing time for a single frame ranged from 50 to 100 ms per frame.

A dataset containing six food crops and eight weed species was used by Sudars et al. [34]. The dataset size is 7,853. To introduce variability, the food crops and weeds were grown both in the field and in a controlled environment. The dataset was manually annotated by an expert. Three RGB digital color cameras were used to capture images of crops and weeds at different growth stages. The image resolutions used for capturing food crops and weed species images were 640×360×3, 720×1,280×3, 480×384×3, 640×480×3, and 1,000×750×3 pixels. This dataset is available online (open-access) and can be utilized for training DL models for food crops and weed species identification. Jiang et al. [35] conducted an experiment to identify crops and weeds using a CNN-based GCN and ResNet-101. The dataset consists of mixed weed, corn, lettuce, and radish. The DL model achieved a classification accuracy of 97.80% for corn, 99.37% for lettuce, 98.93% for radish, and 96.51% for the mixed weed class. Espejo-Garcia et al. [36] employed a DL network as a feature extractor and a traditional ML algorithm as a classifier. DL CNN networks such as DenseNet, Vignette's, MobileNet, Inception-Resnet, and Xception were used as feature extractors. The extracted features from the DL models were then used as input for traditional ML models including SVM, LR, and XGBoost. Among all combinations of networks and classification algorithms, DenseNet + SVM exhibited the highest classification accuracy. The F1-score was found to be 99.29%.

9.5 PLANT DISEASE DETECTION

Crop yield heavily relies on proper disease management. The uniform spraying of agrochemicals over the cropping area is the most widely used method, but it results in significant chemical wastage, high costs, and environmental damage in both open-field and covered cultivation. Farmers and experienced individuals have traditionally relied on their naked eyes to recognize crop diseases based on their experience and traditional wisdom. However, this visual disease scanning method is time-consuming and labor-intensive, making it impractical for large farms. For plant disease detection, modern techniques such as traditional ML and CNN-based DL classification methods outperform older methods (expert judgment) in terms of time savings and detection accuracy [37].

9.5.1 TRADITIONAL MACHINE LEARNING (ML) MODELS

The color feature was used to classify diseased and healthy plants [38]. The RGB color space was transformed to YCbCr and CIELAB. Image smoothing was performed using a Median filter. The OTSU thresholding technique was used for image segmentation. The only remaining problem is distinguishing the same-colored veins from spots. Authors proposed a diseases detection model for Brinjal crop leaves. An unsupervised machine algorithm called K-means clustering algorithm was used for background pixel removal. Gabor filter was used to extract features from healthy and unhealthy Brinjal crop leaves. ANN was used for the classification of healthy and unhealthy plant leaves of Brinjal.

A model was developed for detecting the presence of illness in potato plant leaves that combines ML and image processing [39]. Training and testing were done on images obtained from the 'Plant Village' dataset. Before proceeding to training and testing, the image background was removed, and a mask image was created. Color and texture features were extracted from the masked images and these features were used as input to ML models. Texture features from the images were extracted using the GLCM method. SVM was used as a classifier. The accuracy of the disease detection model was around 95% when tested on 300 images.

GLCM texture analysis was employed by Mahmud et al. [21] to detect strawberry powdery mildew disease under natural and artificial cloud lighting circumstances. After converting the color image to an HSI (Hue, Saturation, and Intensity) image, 10 characteristics were retrieved from the HSI color channel. The angles of offset and orientation were 1? and 0?, respectively. Texture features of diseased and healthy plants were extracted from four channels (green ratio, hue, saturation, and intensity) using a GLCM. A total of 50 texture features were recovered from photos; however, stepwise discriminant analysis was used to choose 25 features for illness identification. Discriminant function analysis was used for the classification of diseased and healthy plants. Mahmud et al. [40] used GLCM texture analysis for feature extraction from strawberry powdery mildew disease images. They also used the ANN technique to process images and classify them as disease and non-disease images. The researchers employed a back-propagation ANN with a tanh sigmoid activation function and a 15,000-epoch size. Field site-I had the highest recall, precision, and F-measure, respectively, of 82.09%, 87.65%, and 85.09%.

Field sites-II and III had the lowest recall, precision, and F-measure. Due to high wind speeds (>8 km/h), the presence of another illness (spider mite), leaf tilt, and overlap, a small divergence in system performance was found. A HSI sensor was used for acquiring images of sugar beet diseases (healthy and unhealthy) [41]. This study was conducted to identify sugar beet disease. The SVM algorithm and 11 spectral vegetation indices were used for the classification of sugar beet disease. The classification accuracy of the model was reported as 97% in classifying images into healthy and unhealthy.

9.5.2 CNN-BASED DEEP LEARNING (DL) MODELS

A smartphone-assisted disease detection system has been developed [42]. The study was primarily motivated by the increasing global usage of smartphones and recent advancements in computer vision through DL. A deep CNN was trained to identify 14 crop species and 26 diseases using a public dataset of 54,306 photos of healthy and damaged plant leaves, captured under controlled conditions. These images were obtained from the 'Plant Village' dataset and resized to 256 x 256 pixels. Various combinations of training and testing splits were explored. Two approaches were followed in developing the DL models. The first approach involved using pre-trained transfer learning models for classification model development. In the second approach, a CNN-based model for classification was built from scratch. The impact of image type on classification accuracy was investigated, with color images, grayscale images, and background-removed images used as inputs to the network. The accuracy of the two models was evaluated using the F1-score. The average F1-score for GoogleLeNet and AlexNet was 98.86% and 98.48%, respectively. The classification accuracy of the model reached up to 99.35%.

A CNN-based DL network was developed to classify tubers into five classes, namely healthy and four disease classes [43]. The dataset contains potatoes of different shapes, sizes, and tones acquired under normal conditions. The dataset was manually labeled by a subject matter specialist. The total dataset was divided into various combinations of train and test splits to observe the impact on model accuracy. The highest classification accuracy achieved was 96% with a training dataset size of 90% and a testing dataset size of 10%. GoogleLeNet and Inception v3 models were implemented on TensorFlow to detect two types of pests and three diseases in Cassava crops [44]. The dataset consisted of 11,670 images used for training, validation, and testing stages. The performance metric used was the confusion matrix. The classification accuracy ranged from 80% to 93%.

Waheed et al. [45] utilized an optimized DenseNet DL model for the detection and classification of three corn leaf diseases. The total number of trainable parameters was 77,612. The optimized DenseNet's classification accuracy was compared with VGGNet, XceptionNet, EfficientNet, and NASNet. The DL models were trained on a dataset of 12,332 images, each with an image resolution of 250×250×3. Data augmentation techniques such as cropping, padding, and horizontal flipping were employed to increase the relevant data. The accuracy of the DenseNet model was found to be 98.06%. In the future, the researchers intend to develop a mobile application for corn leaf disease detection and classification.

AlexNet, VGG16, and VGG19 CNN models were utilized for the detection of diseases in rice fields [46]. Three models were trained to classify four types of paddy diseases, namely Brown spot, Bacterial blight, Tungro, and Blast. The SVM classifier was employed for the detection of these four types of rice diseases. In this study, CNN models utilized a feature extractor from 5,932 disease images. In the second stage, the extracted features were inputted into the SVM classifier. AlexNet (feature layer: fc8) and ResNet50 with the SVM algorithm exhibited superior classification performance compared to other algorithms, achieving an F1-score of 98.38%. A CNN-based DL model was trained with 10,000 labeled images for the detection of Cassava crop diseases [47]. The images were processed using the adaptive histogram equalization technique. The accuracy of the model ranged from 76.9% to 99.30%. A CNN and ANN model were employed for the detection of plant diseases [48]. Feature extraction with CNN and ANN classification yielded better results than other supervised ML models under different positions or leaf angles in field conditions.

9.6 FACTORS AFFECTING ON DISEASE AND WEED DETECTION USING ML AND DL TECHNIQUES

In comparison to laboratory conditions, imaging in the open field is a challenging task. Texture and shape features have shown promising results in classification problems, but their performance is affected by factors such as high wind [21], overlapping, and high weed density. Crop factors, including leaf angle, leaf overlapping [26], and leaf density, also play a role. The overlapping of leaves and wind speed can impact the performance of classifiers in real field conditions. Calculating shape features becomes difficult when leaves are overlapped or damaged by insects. Therefore, considering texture features along with shape features can improve the performance of ML models. Illumination intensity varies throughout the day [49], and wind speed is a major concern in the field. Uncontrolled outdoor illumination can affect the performance of imaging sensors. The accuracy and precision of classification models are influenced by the quality of input images. Artificial lighting conditions and protective cloth can help safeguard the performance of imaging sensors [21]. Weed detection and classification can be enhanced using computer vision-based ML and DL algorithms, without any adverse environmental effects. These advancements in science and technology will empower robots to detect pests and apply insecticides to diseased plants in the future. Furthermore, robots will be able to apply herbicides in a site-specific manner, minimizing the need for herbicides.

9.7 ADVANTAGES AND DISADVANTAGES OF DISEASE AND WEED DETECTION USING MACHINE LEARNING (ML) AND DEEP LEARNING (DL) TECHNIQUES

The advantage of traditional ML and DL-based classification is that it is simpler and faster than traditional scanning techniques or manual classification techniques. The main advantage of traditional ML-based classification or regression is that a small amount of data is needed for training the model compared to DL models. Additionally, the training time for ML models is shorter and does not require high GPU for computation. DL models can be converted into TensorFlow Lite and deployed on any microcontroller, such as Raspberry Pi. Furthermore, spraying units can be attached to the microcontroller to achieve site-specific herbicide or pesticide application. The disadvantage of DL is that it requires a large dataset to develop DL models. Generating data is time-consuming, and environmental factors such as rain, clouds, light, and wind can affect the data collection process. DL models are computationally intensive and require GPU and TPU.

KEYWORDS

- **agriculture**
- **color spaces**
- **convolutional neural network**
- **deep learning**
- **disease detection**
- **image processing**
- **machine learning**
- **TensorFlow**
- **weeds detection**

REFERENCES

1. Sharma, A., Jain, A., Gupta, P., & Chowdary, V., (2020). Machine learning applications for precision agriculture: A comprehensive review. *IEEE Access*.
2. Liu, L., Dong, Y., Huang, W., Du, X., Ren, B., Huang, L., ZHENG, Q., & Ma, H., (2020). A Disease index for efficiently detecting wheat fusarium head blight using sentinel-2 multispectral imagery. *IEEE Access*, *8*, 52181–52191.
3. Singh, V., Sharma, N., & Singh, S., (2020). A review of imaging techniques for plant disease detection. *Artificial Intelligence in Agriculture, 4*, 229–242.
4. Liu, B., & Bruch, R., (2020). Weed detection for selective spraying: A review. *Current Robotics Reports*, *1*(1), 19–26.
5. Wu, Z., Chen, Y., Zhao, B., Kang, X., & Ding, Y., (2021). Review of weed detection methods based on computer vision. *Sensors*, *21*(11), 3647.
6. Hamuda, E., Glavin, M., & Jones, E., (2016). A survey of image processing techniques for plant extraction and segmentation in the field. *Computers and Electronics in Agriculture*, *125*, 184–199.
7. Agrawal, K. N., Singh, K., Bora, G. C., & Lin, D., (2012). Weed recognition using image-processing technique based on leaf parameters. *Journal of Agricultural Science and Technology. B*, *2*(8B), 899.
8. Deng, L. M., Tang, J., & Ma, W. J., (2014). Feature extraction and recognition system of maize leaf based on image processing. *Chinese Journal of Mechanical Engineering, 35*, 72–75.
9. Dryden, I. L., Scarr, M. R., & Taylor, C. C., (2003). Bayesian texture segmentation of weed and crop images using reversible jump Markov chain Monte Carlo methods. *Journal of the Royal Statistical Society: Series C (Applied Statistics)*, *52*(1), 31–50.

10. Jinglei, T., Ronghui, M., Zhiyong, Z., Jing, X., & Dong, W., (2017). Distance-based separability criterion of ROI in classification of farmland hyper-spectral images. *International Journal of Agricultural and Biological Engineering, 10*(5), 177–185.
11. Slaughter, D. C., Giles, D. K., & Downey, D., (2008). Autonomous robotic weed control systems: A review. *Computers and Electronics in Agriculture, 61*(1), 63–78.
12. Longchamps, L., Panneton, B., Samson, G., Leroux, G. D., & Thériault, R., (2010). Discrimination of corn, grasses, and dicot weeds by their UV-induced fluorescence spectral signature. *Precision agriculture, 11*(2), 181–197.
13. Huang, Y., Lee, M. A., Thomson, S. J., & Reddy, K. N., (2016). Ground-based hyperspectral remote sensing for weed management in crop production. *International Journal of Agricultural and Biological Engineering, 9*(2), 98–109.
14. Shapira, U., Herrmann, I., Karnieli, A., & Bonfil, D. J., (2013). Field spectroscopy for weed detection in wheat and chickpea fields. *International Journal of Remote Sensing, 34*(17), 6094–6108.
15. Meng, Q., Qiu, R., He, J., Zhang, M., Ma, X., & Liu, G., (2015). Development of agricultural implement system based on machine vision and fuzzy control. *Computers and Electronics in Agriculture, 112*, 128–138.
16. Yu, J., Schumann, A. W., Cao, Z., Sharpe, S. M., & Boyd, N. S., (2019). Weed detection in perennial ryegrass with deep learning convolutional neural network. *Frontiers in Plant Science, 10*, 1422.
17. Lin, T. Y., Maire, M., Belongie, S., Hays, J., Perona, P., Ramanan, D., & Zitnick, C. L., (2014). Microsoft coco: Common objects in context. In: *European Conference on Computer Vision* (pp. 740–755). Springer, Cham.
18. Kuznetsova, A., Rom, H., Alldrin, N., Uijlings, J., Krasin, I., Pont-Tuset, J., & Ferrari, V., (2020). The open images dataset v4. *International Journal of Computer Vision, 128*(7), 1956–1981.
19. Deng, J., Dong, W., Socher, R., Li, L. J., Li, K., & Fei-Fei, L., (2009). ImageNet: A large-scale hierarchical image database. In: *2009 IEEE Conference on Computer Vision and Pattern Recognition* (pp. 248–255). IEEE.
20. Everingham, M., Van, G. L., Williams, C. K., Winn, J., & Zisserman, A., (2010). The Pascal visual object classes (VOC) challenge. *International Journal of Computer Vision, 88*(2), 303–338.
21. Mahmud, M. S., Zaman, Q. U., Esau, T. J., Price, G. W., & Prithiviraj, B., (2019). Development of an artificial cloud lighting condition system using machine vision for strawberry powdery mildew disease detection. *Computers and Electronics in Agriculture, 158*, 219–225.
22. Chandel, N. S., Chakraborty, S. K., Rajwade, Y. A., Dubey, K., Tiwari, M. K., & Jat, D., (2021). Identifying crop water stress using deep learning models. *Neural Computing and Applications, 33*(10), 5353–5367.
23. Chang, Y. K., Zaman, Q. U., Schumann, A. W., Percival, D. C., Esau, T. J., & Ayalew, G., (2012). Development of color co-occurrence matrix-based machine vision algorithms for wild blueberry fields. *Applied Engineering in Agriculture, 28*(3), 315–323.
24. Rehman, T. U., Zaman, Q. U., Chang, Y. K., Schumann, A. W., Corscadden, K. W., & Esau, T. J., (2018). Optimizing the parameters influencing performance and weed

(goldenrod) identification accuracy of color co-occurrence matrices. *Biosystems Engineering, 170*, 85–95.
25. Perez, A. J., Lopez, F., Benlloch, J. V., & Christensen, S., (2000). Color and shape analysis techniques for weed detection in cereal fields. *Computers and Electronics in Agriculture, 25*(3), 197–212.
26. Utstumo, T., Urdal, F., Brevik, A., Dørum, J., Netland, J., Overskeid, Ø., & Gravdahl, J. T., (2018). Robotic in-row weed control in vegetables. *Computers and Electronics in Agriculture, 154*, 36–45.
27. Zhu, W., & Zhu, X., (2009). The application of support vector machine in weed classification. In: *2009 IEEE International Conference on Intelligent Computing and Intelligent Systems* (Vol. 4, pp. 532–536). IEEE.
28. He, D., Qiao, Y., Li, P., Gao, Z., Li, H., & Tang, J., (2013). Weed recognition based on SVM-DS multi-feature fusion. *Nongye Jixie Xuebao= Transactions of the Chinese Society for Agricultural Machinery, 44*(2), 182–187.
29. Bakhshipour, A., Jafari, A., Nassiri, S. M., & Zare, D., (2017). Weed segmentation using texture features extracted from wavelet sub-images. *Biosystems Engineering, 157*, 1–12.
30. Bakhshipour, A., & Jafari, A., (2018). Evaluation of support vector machine and artificial neural networks in weed detection using shape features. *Computers and Electronics in Agriculture, 145*, 153–160.
31. Le, V. N. T., Apopei, B., & Alameh, K., (2019). Effective plant discrimination based on the combination of local binary pattern operators and multiclass support vector machine methods. *Information Processing in Agriculture, 6*(1), 116–131.
32. Chen, Y., Wu, Z., Zhao, B., Fan, C., & Shi, S., (2021). Weed and corn seedling detection in field based on multi-feature fusion and support vector machine. *Sensors, 21*(1), 212.
33. Chechliński, Ł., Siemiątkowska, B., & Majewski, M., (2019). A system for weeds and crops identification—reaching over 10 FPS on Raspberry Pi with the usage of MobileNets, DenseNet and custom modifications. *Sensors, 19*(17), 3787.
34. Sudars, K., Jasko, J., Namatevs, I., Ozola, L., & Badaukis, N., (2020). Dataset of annotated food crops and weed images for robotic computer vision control. *Data in Brief, 31*, 105833.
35. Jiang, H., Zhang, C., Qiao, Y., Zhang, Z., Zhang, W., & Song, C., (2020). CNN feature-based graph convolutional network for weed and crop recognition in smart farming. *Computers and Electronics in Agriculture, 174*, 105450.
36. Espejo-Garcia, B., Mylonas, N., Athanasakos, L., Fountas, S., & Vasilakoglou, I., (2020). Towards weeds identification assistance through transfer learning. *Computers and Electronics in Agriculture, 171*, 105306.
37. Nagaraju, M., & Chawla, P., (2020). Systematic review of deep learning techniques in plant disease detection. *International Journal of System Assurance Engineering and Management, 11*(3), 547–560.
38. Chaudhary, P., Chaudhari, A. K., Cheeran, A. N., & Godara, S., (2012). Color transform-based approach for disease spot detection on plant leaf. *International Journal of Computer Science and Telecommunications, 3*(6), 65–70.
39. Islam, M., Dinh, A., Wahid, K., & Bhowmik, P., (2017). Detection of potato diseases using image segmentation and multiclass support vector machine. In: *2017 IEEE 30th*

Canadian Conference on Electrical and Computer Engineering (CCECE) (pp. 1–4). IEEE.

40. Mahmud, M. S., Zaman, Q. U., Esau, T. J., Chang, Y. K., Price, G. W., & Prithiviraj, B., (2020). Real-time detection of strawberry powdery mildew disease using a mobile machine vision system. *Agronomy, 10*(7), 1027.
41. Rumpf, T., Mahlein, A. K., Steiner, U., Oerke, E. C., Dehne, H. W., & Plümer, L., (2010). Early detection and classification of plant diseases with support vector machines based on hyperspectral reflectance. *Computers and Electronics in Agriculture*, *74*(1), 91–99.
42. Mohanty, S. P., Hughes, D. P., & Salathé, M., (2016). Using deep learning for image-based plant disease detection. *Frontiers in Plant Science*, *7*, 1–10.
43. Oppenheim, D., & Shani, G., (2017). Potato disease classification using convolution neural networks. *Advances in Animal Biosciences*, *8*(2), 244–249.
44. Ramcharan, A., Baranowski, K., McCloskey, P., Ahmed, B., Legg, J., & Hughes, D. P., (2017). Deep learning for image-based cassava disease detection. *Frontiers in Plant Science*, *8*, 1852.
45. Waheed, A., Goyal, M., Gupta, D., Khanna, A., Hassanien, A. E., & Pandey, H. M., (2020). An optimized dense convolutional neural network model for disease recognition and classification in corn leaf. *Computers and Electronics in Agriculture*, *175*, 105456.
46. Sethy, P. K., Barpanda, N. K., Rath, A. K., & Behera, S. K., (2020). Deep feature-based rice leaf disease identification using support vector machine. *Computers and Electronics in Agriculture*, *175*, 105527.
47. Sambasivam, G., & Opiyo, G. D., (2021). A predictive machine learning application in agriculture: Cassava disease detection and classification with imbalanced dataset using convolutional neural networks. *Egyptian Informatics Journal*, *22*(1), 27–34.
48. Shin, J., Chang, Y. K., Heung, B., Nguyen-Quang, T., Price, G. W., & Al-Mallahi, A., (2020). Effect of directional augmentation using supervised machine learning technologies: A case study of strawberry powdery mildew detection. *Biosystems Engineering*, *194*, 49–60.
49. Kazmi, W., Garcia-Ruiz, F. J., Nielsen, J., Rasmussen, J., & Andersen, H. J., (2015). Detecting creeping thistle in sugar beet fields using vegetation indices. *Computers and Electronics in Agriculture*, *112*, 10–19.

PART IV

Artificial Intelligence and Machine Learning for Security Solutions

CHAPTER 10

VIDEO IMAGE IDENTIFICATION AND VERIFICATION USING PRELIMINARY SCORE APPROACH FOR AUTHENTICATION AND SECURITY ENHANCEMENT

ADITI UPADHYAY,[1] SHRIKANT UPADHYAY,[2] and SUDHIR KUMAR SHARMA[1]

[1]*Jaipur National University, Jaipur, Rajasthan, India*

[2]*Cambridge Institute of Technology, Bangalore, Karnataka, India*

ABSTRACT

Image is one of the identities of a person, which reflects their emotions, feelings, age, etc. It helps gather information about a person without knowing their name, caste, or age. This may not be of much importance when used for domestic or framing applications. However, when images are involved in multimedia, privacy becomes a major concern for everyone. Everyone wants to receive the original image without it being tampered with by unauthorized individuals. Image authentication has gained much attention due to its wide usage in multimedia and various applications. The non-secure transmission of digital images over channels is increasing globally day by day through computer networks, making the authentication of each image a challenging task. Therefore, defense, space, medical, and quality measurement images must be secured against wrongful manipulation or changes to the original image. To ensure the authenticity of multimedia pictures, various methods

The Fusion of Artificial Intelligence and Soft Computing Techniques for Cybersecurity.
M. A. Jabbar, Sanju Tiwari, Subhendu Kumar Pani, & Stephen Huang (Eds.)

have been proposed, including digital signatures, watermarking, cryptography, and fragility based on image contents. The objective of this chapter is to identify and verify real-time video images using a preliminary score approach. This approach helps to assess the authenticity of an image at an initial stage by extracting features, evaluating them using the Viterbi algorithm, and converting the input image into an embedded state. The achieved matrix is then transformed, and based on this, a preliminary score estimation is generated after multiple iterations for each image. Finally, the tested image is verified using the aforementioned approaches to protect and provide security to the original image. This approach can be useful in various surveillance applications for real-time image identification and verification.

10.1 INTRODUCTION

In the past few years, face identification and its verification have received great attention in the fields of computer vision, pattern recognition, and biometrics. Researchers are particularly interested in the remarkable ability to identify individuals, considering the relevance of human activity in both daily life and multimedia. Facial analysis involves extracting or gathering information such as age, gender, pose, landmarks, and identity. It has various applications, including law enforcement, face biometrics for transactions, self-driving vehicles, and active verification on devices [1–6].

Face identification and verification systems typically consist of three stages. The first stage requires a face identifier to locate faces in an image, which should possess strong properties to handle variations in illumination, scale, and pose. Additionally, an efficient face identifier should consistently produce accurate results while confining the area of interest [7]. The second stage localizes facial indicators such as the tip of the nose, eye centers, tip of the ear lobes, and corner of the mouth. These indicators are then used to align faces, taking into account scaling and in-plane rotation effects. The third stage involves a feature extractor that encodes identity statistics into a high-dimensional locator. These comprehensive descriptions are then used to evaluate similarity or score between similar faces. A robust feature extractor is capable of handling errors introduced by previous stages in the pipeline: face alignment, face detection, and landmark localization [8]. The steps involved in image identification and verification are shown in Figure 10.1. The face has many advantages that make it an efficient biometric for distinguishing individuals. First, the face reflects not only the identity but also the internal emotions (feelings) of happiness or sadness, including the approximate age of the person. Second, even at a long distance, capturing the face biometric is easy. This feature makes face recognition important in person recognition as well as in human-computer interaction [9].

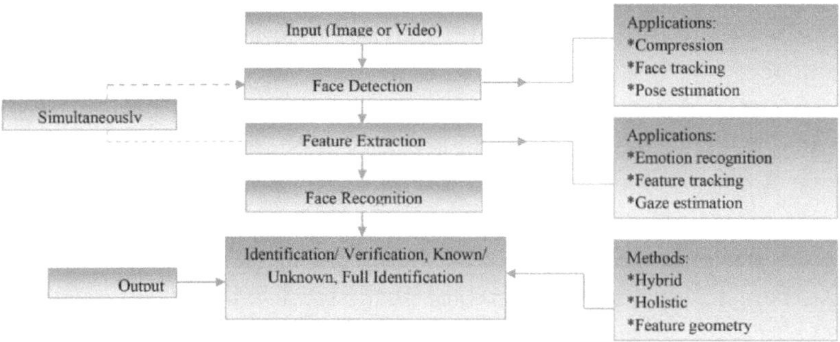

FIGURE 10.1 Steps involved in identification and verification.

Source: Reproduced with permission from Upadhyay et al., 2023. https://creativecommons.org/licenses/by/4.0/

The face biometric is affected by a number of extrinsic and intrinsic variations such as lighting, pose, age, and expressions. Over the past decades, there has been a significant improvement in the performance of face recognition, but it still remains below acceptable levels in many applications [10–16]. Recent efforts have mainly focused on different features, 3D (three-dimensional) models, and video inputs to overcome the performance limitations of 2D still face recognition. Geometrically, the face is essentially a 3D space composed of a large number of polygons that can be represented by facial features and pixels related to the face geometry. Image data can be computationally represented in vector form. For example, an $x \times y$ 2D image can be mapped to a vector $P \varepsilon R^{xy}$ by sequencing the pixel elements through concatenating each row and column of the image. Different faces with different emotions are shown in Figure 10.2. Where; $P=(P_1,P_2 \ldots P_i \ldots P_N)$, denotes the $n \times N$ data matrix; where; P_i is a face vector of dimension n, concatenated from $x \times y$ face image, where $n=x \times y$ denotes the total number of pixels.

FIGURE 10.2 Humans faces with emotions.

Source: Reproduced with permission from Upadhyay et al., 2023. https://creativecommons.org/licenses/by/4.0/

10.2 PRELIMINARY SCORE APPROACH

Face identification at the preliminary stage will be done using the hidden Markov model (HMM) with unobserved states. These states do not directly reflect the identity of the face but generate tokens in the form of sequences that highlight the information of the sequence state. Variable $\alpha(t)=\{\alpha_1,\alpha_2,\alpha_3\}$ is the state (hidden) with time 't' and variable $b(t)=\{b_1,b_2,b_3\}$ at time 't.' Now, using gray scale for 1D HMM, the facial area covers the hairs, forehead, eyes, nose, and mouth in an up-to-down order. The face recognition system is shown in Figure 10.3. It is observed that each facial area is assigned to a state from left to right in 1D. However, embedded HMMs in 2D have been utilized for image representation and recognition, $Y=\{Y_t\}=\{Y_{tpq}\}$ using spatial indices (p,q) and designed by HMM arrangement which consist of a set called super state S_{pq} which does not allow any type of crossing between embedded state and super state [17–23]. A fully comprehensive state includes five characteristics that properly cover the forehead, eyes, nose, mouth, and chin in a vertical direction, along with initial state and state transition probabilities. $\{\Pi v, B_v\}=\{\pi_{fn};b_{f_{nFn+1}}\}$. Now, every state (super) itself is a 1D standard HMM containing an embedded state for facial modeling features and has 21 embedded states in the horizontal direction.

$$b_{pqr}(x)=p\left(x\,\mathrm{I}\,\mathrm{r}_{pqr},p_b\right)$$

$$=\acute{\mathrm{O}}_{1_{pq}}^{N_1}\mathrm{w}_{pq}\mathrm{r}1_{pq}\mathrm{N}(\mathrm{x}\,\mathrm{I}\grave{\mathrm{i}}_{pqr}1_{pq},\acute{\mathrm{O}}_{pqr}1_{pq}) \quad (1)$$

The distribution of two different videos to attain values of different score levels is shown in Figure 10.4(a) and (b).

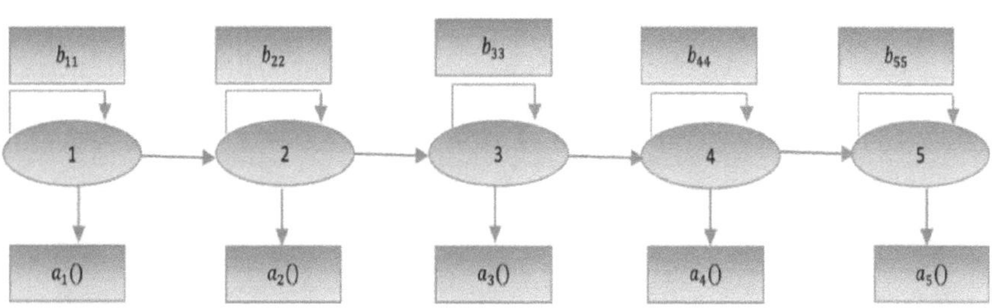

FIGURE 10.3 HMM for face recognition (left-to-right).

Source: Reproduced with permission from Upadhyay et al., 2023. https://creativecommons.org/licenses/by/4.0/

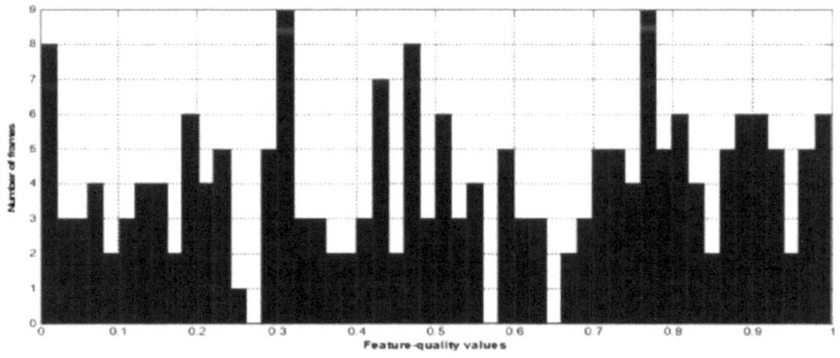

(a)

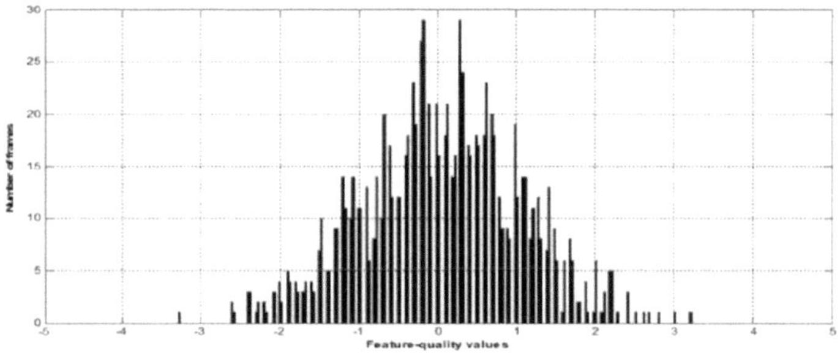

(b)

FIGURE 10.4 Feature distribution for two dissimilar videos ((a) and (b)) where some values reach zero, while other features attain values close to one. This indicates that high fidelity frames have the maximum score, while poor frames have the minimum score. Here, the x-axis represents feature quality values, and the y-axis represents the number of frames.

Source: Reproduced with permission from Upadhyay et al., 2023. https://creativecommons.org/licenses/by/4.0/

10.3 IMAGE AUTHENTICATION

Authentication means verifying the real value or content, which is achieved without being tampered with. In real-time applications, it is very difficult to maintain the original identity due to various reasons generated by the surroundings or communication devices. Image authentication does not accept any tampering in the image data. The conventional approach based on cryptography evaluates message authentication from images using a hash function. This hash function is then encrypted with the sender's private key and attached to the image. For more secure data exchange between parties, the hash can be encrypted with the help of the beneficiary's public key. The authentication and verification process are depicted in

Figure 10.5(a) and (b). The recipient evaluates the hash from the received image, which was attached to the recovered image and decrypted with the help of the private key. Hash evaluation of images using column and line functions is obtained, and these hashes are stored and compared with those obtained from each column and line of the image to be verified. If any mismatch in the hash function is observed, the image is declared to be tampered with; otherwise, it is announced as authentic. Since hash functions are very sensitive to any slight changes in binary data or image pixels, this puts limits on image verification for many applications. The fragile watermarking approach is used to create a watermark and place it in the image to be protected in such a way that any changes found are also reflected in the placed watermark. Verifying the existence of the placed watermark approves the authenticity of the image and finally localizes the damaged areas. This method does not accept any distorted image and authenticates only if all its pixels remain constant. The generation and verification process are depicted in Figure 10.6(a) and (b).

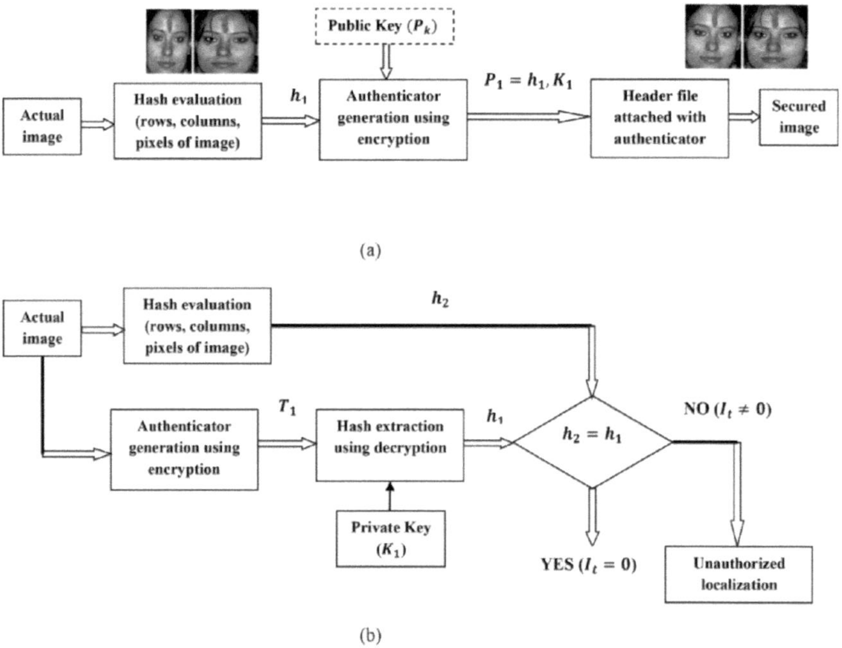

FIGURE 10.5 System for authentication using cryptography – (a) authenticator generator; and (b) authenticity verification.

Source: Reproduced with permission from Upadhyay et al., 2023. https://creativecommons.org/licenses/by/4.0/

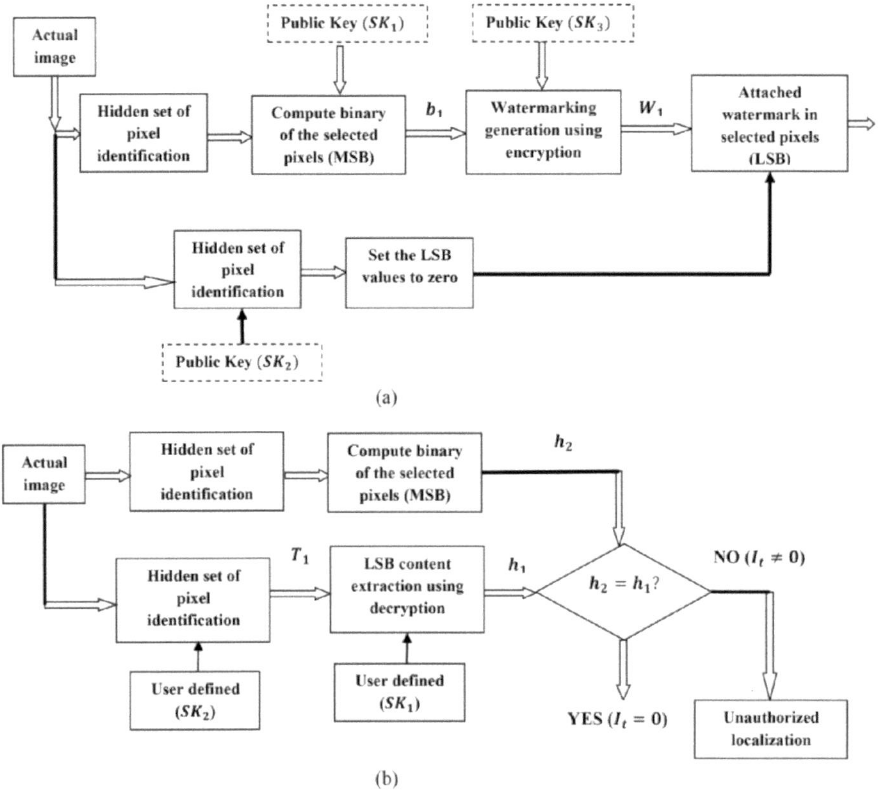

FIGURE 10.6 Authentication using fragile watermarking – (a) authenticator generator; and (b) authenticity verification.

Source: Reproduced with permission from Upadhyay et al., 2023. https://creativecommons.org/licenses/by/4.0/

Fragile watermarking algorithms are based on forming a watermark from the identity of an image, and the watermark is evaluated from a set of image pixels. The set of pixels, taken from the secret key SK_1, can be encrypted with another set's least significant bits using the secret key SK_3. To enhance the security of the algorithm, the set of pixels can be determined by another secret key, SK_2.

The purpose of a digital signature is to electronically sign a document, such as when signing a contract or withdrawing money from a bank. Digital signatures can be transferred electronically along with a signed letter. However, there are several issues associated with digital signatures.

Firstly, a digital signature requires an algorithm to be attached to the signature for electronic transfer. Secondly, there are problems related to verification and signature authentication. For example, when a person's signature is verified by comparing it with their certified signature while making a purchase using a credit/debit card. This authentication approach is not strong enough, as it is easy to copy the signature of a certified person. Therefore, electronic signatures can be tested by anyone with knowledge of the verification algorithm.

Applying the digital signature algorithm directly to an image may result in situations where the image content has not been tampered with, but the algorithm declares the image as non-authentic. As a result, changes to existing digital signature methods must be made by defining the information that needs to be signed, such as signing the image content rather than the image data itself. Digitally signing the hash value using existing digital signatures like a public key or private key can increase overall security. The verification procedure depends on the image signature, which can be extracted from the attached signature and compared with the original signature [24–30].

A hash function operates on message '*X*' of any length to achieve a fixed-size hash '*h*' called a message digest.

$$h=H(X) \tag{2}$$

where; H indicates hash function and size of h is usually smaller than 'X.'

The digital signature algorithm and hash function have the same features, which are used to compare the signature based on dissimilarity and selected characteristics. A variable length code can also be assigned to the digital signature for verification. Coefficient quantization is applied to obtain a {0,1} sequence, which can be encoded to form the digital signature. This algorithm is valid with 2 and 3 decomposition levels, but there is a trade-off between complexity, security, and processing time. Cryptography provides higher-level solutions for digital image security. Conventional cryptography provides strict authentication rules that determine the authenticity of an image based on its binary representation or changes in image pixels. This approach has good performance, but the localization of the evaluated areas is not up to the mark. The detection of tampered images was done well using the conventional cryptography algorithm approach. The subset of frames represented in Figure 10.7 under different poses, illuminations, and may be useful for identification purposes.

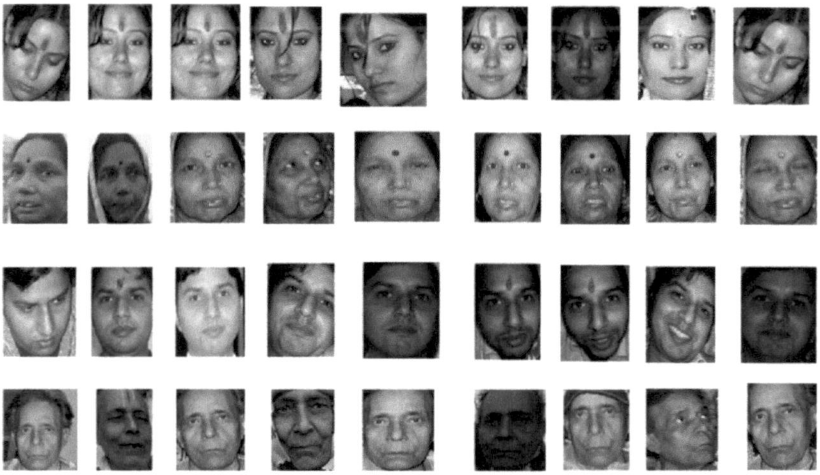

FIGURE 10.7 Subset of frames reflecting the information present in a video. A single video captures images under different illuminations, poses, and expressions from the same occasion. Some frames prove to be useful for face identification, while others can be detrimental to performance.

Source: Reproduced with permission from Upadhyay et al., 2023. https://creativecommons.org/licenses/by/4.0/

Video face verification requires matching all the frames in two videos. However, not all frames are equally informative and may have extreme variations in pose, image quality, illumination, and expression. Therefore, some frames may be damaged due to the presence of covariates, making it crucial to select and utilize video information efficiently and carefully.

10.4 ALGORITHM FOR FRAME SELECTION AND AVERAGE SCORE

The algorithm for video image frame selection and its estimated coefficients for maximum average score are discussed in the following steps:

1. Video frame selection is based on the average using the discrete wavelet transform method, which first identifies the given input image *(I)* that can be evaluated as:

$$I^{\mathrm{I}}(DWT) = \left[I^{\mathrm{I}}{}_{ap}, I^{\mathrm{I}}{}_{v,h,d}\right]$$

where; I_{ap} represents the estimated coefficients of the image; and $I_{v,h,d}$ contains coefficient details of vertical, horizontal, and verticals sidebands.

2. A filtering process will be applied using high pass and low pass filters for the decomposition of the parent wavelets.
3. The next level of DWT has now been applied to the first approximation achieved in the first step approximation band.

$$I^{I}(DWT) = \left[I^{I}{}_{ap}, I^{I}{}_{v,h,d}\right]$$

where; $I^{I}{}_{ap}$, $I^{I}{}_{v,h,d}$ denotes second level estimation of input image (I).

4. The average of every DWT band is evaluated by dividing the image into x×y windows and capturing them.

$$H(j) = -\sum_{i=1}^{m} P(j_i)\log_2 P(j_i)$$

where; $P(j_i)$ is the probability mass function that indicates pixel values in probability form reflecting in the neighborhood; and m is the total number of pixels.

5. If window size is $L_x \times W_y$ then:

$$P(j_i) = \frac{m_{ji}}{L_x \times W_y}$$

where; m_{ji} denotes number of pixels in windows.

6. The average of each window is now combined to evaluate the feature value of the band.

$$FH = \sum_{i=1}^{\mu}(|E_i|)$$

where; FH indicates the feature score value; p is the number of windows; and E_i is the entropy of the i_{th} window.

7. Final total score of images I,$FH(I)$ was obtained by averaging the feature value of each band individually:

$$FH(I) = FH\left(I^{I}_{ap}\right) + FH\left(I^{I}_{h,v,d}\right) + FH\left(I_{h,v,d}\right)$$

8. For a video image V_i; feature score of frames (f_i) is denoted by $FH(f_i)$ and obtained max-min normalization using:

$$m_{min} = \frac{FH(f_i) - \min(FH)}{\max(FH) - \min(FH)}$$

where; FH denotes all feature scores for V_i; and min(FH), max (FH) denotes the denotes the minimum and maximum value of FH.

9. Once the score of each individual frame is evaluated, an adaptive process for frame selection is performed to identify the best set of frames.

$$\theta_i = \begin{cases} 1, if\ n_i \geq \sigma_n + \dfrac{\mu_n}{2} \\ 0, \text{otherwise} \end{cases}$$

where; μ_n denotes the mean appropriate for the feature value ; and σ_n represents the standard deviation.

10. Testing was performed on the rich feature frame retrieved from the database and verified using the matched score for perfect authentication. The false acceptance and false rejection rates are higher for the feature trained image compared to Eigen, random, and normal feature images. From a security standpoint, it is necessary to keep these rates low in order to achieve a desirable high verification performance. However, the high false acceptance rate of the feature trained image is a positive indication of achieving a high verification rate. This is beneficial in maintaining security measures in real-time image applications such as video surveillance and authentication, as depicted in Figures 10.8 and 10.9.

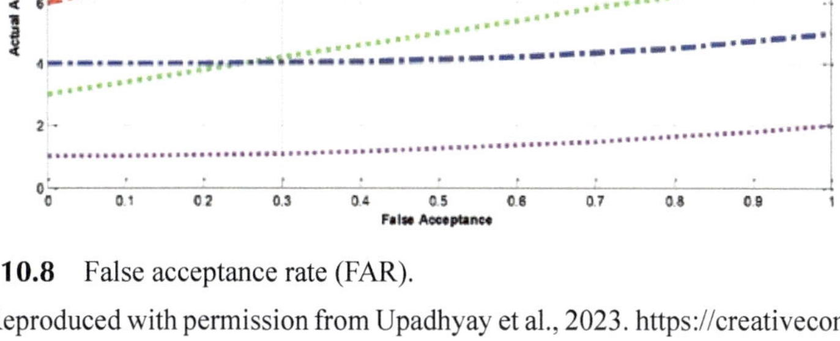

FIGURE 10.8 False acceptance rate (FAR).

Source: Reproduced with permission from Upadhyay et al., 2023. https://creativecommons.org/licenses/by/4.0/

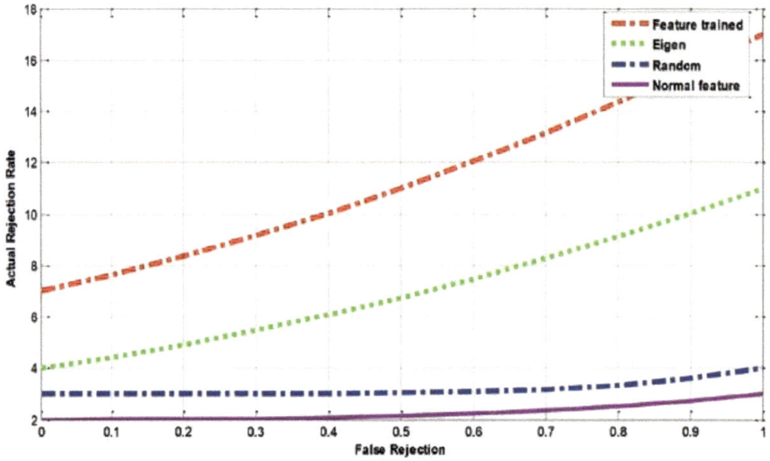

FIGURE 10.9 False rejection rate (FRR).

Source: Reproduced with permission from Upadhyay et al., 2023. https://creativecommons.org/licenses/by/4.0/

10.5 PROPOSED MODEL FOR IDENTIFICATION AND VERIFICATION

The proposed model for face identification and verification is shown in Figure 10.10. It is used to identify the succession of faces from the given input and track the identified face over successive frames. The model selects the most suitable and best face for proper identification, assuming that the background is previously known and fixed. If no face is detected, the system estimates the background by using information from the previous frame. In the final stage of the verification process, if a new face is identified, it is switched to the training stage to create a new face model. If the face is identified from the stored face in the database, the analogous face model is re-trained. Based on the proposed model, identification and verification have been performed on four input images (feature trained, Eigen, random, and random). The verification percentage is calculated by comparing the false acceptance rate of the observed results [31–34]. Real-time images have been captured from various sources such as YouTube videos, social networking sites, and our own database. The results have been simulated to achieve good performance, as discussed in the "Results and Discussions" section.

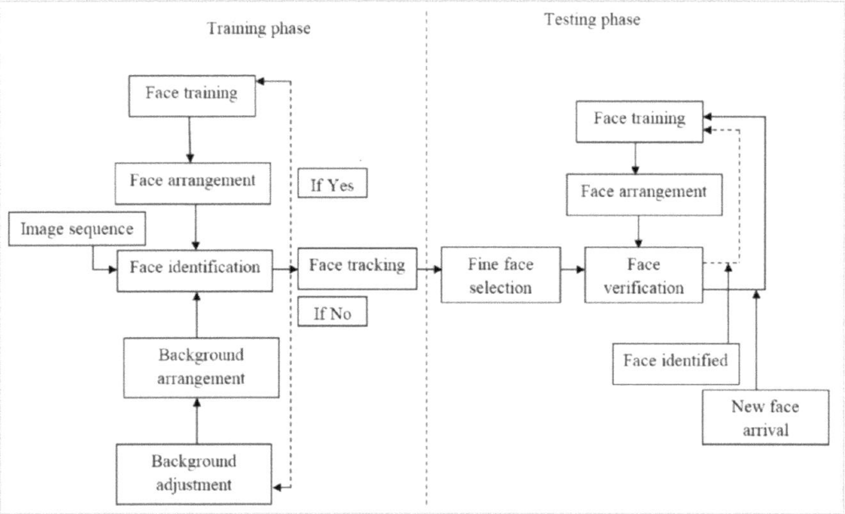

FIGURE 10.10 Proposed model for face identification and verification with training and testing stages.

10.6 RESULTS AND DISCUSSIONS

The experiments are performed with the images gathered from social networking videos, including your database, which contains 10–100 frames with a duration of 2–6 seconds. These frames have informative-based features that were first trained and tested using the proposed model mentioned above for verification percentages. We analyze the performance for a fixed number of frames, without considering adaptive methods or any frame selection algorithms.

The performance report depicted in Figures 10.11–10.14 indicate that the accuracy of feature-trained images for different subjects is more efficient compared to Eigen, random, and normal feature images. Instagram shows higher accuracy compared to other platforms such as Facebook, YouTube, and the self-database images. This may be due to the fact that Instagram images are unaltered and contain all their captured features, whereas YouTube images have lower accuracy compared to other images. Preliminary scores can help evaluate the overall score of each frame, and by averaging the scores at the maximum level, the desired verification rate can be achieved. Security levels can be further improved by assigning each image an unpredictable coding with a private key instead of hash coding (Tables 10.1–10.4).

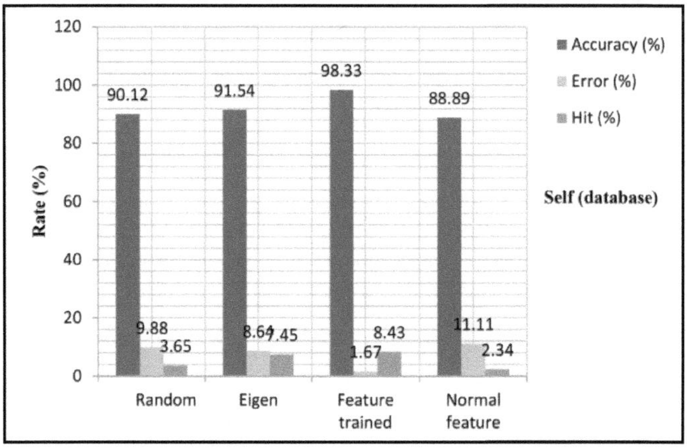

FIGURE 10.11 Self-collected images performance from the video.

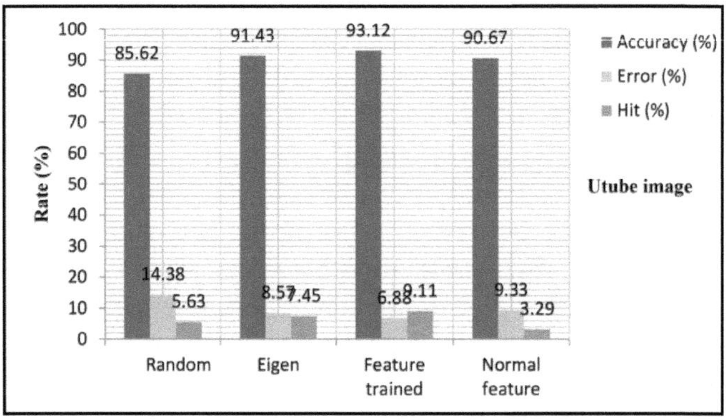

FIGURE 10.12 Performance report for YouTube video images.

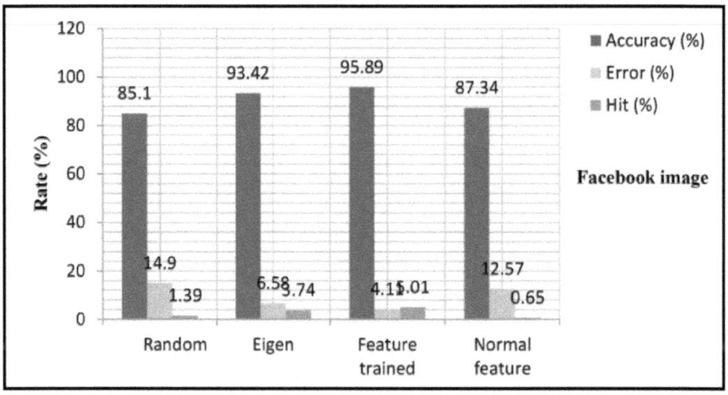

FIGURE 10.13 Performance report for Facebook images.

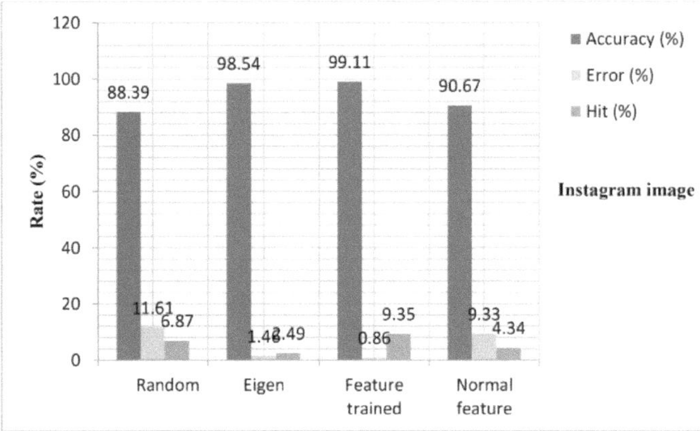

FIGURE 10.14 Performance report for Instagram images.

TABLE 10.1 Database (Self-Created)

Image Type	Identification Accuracy Rate (%)	Verification	
		Error (%)	Hit (%)
Random	90.12	9.88	3.65
Eigen	91.54	8.64	7.45
Feature trained	98.33	1.67	8.43
Normal feature	88.89	11.11	2.34

TABLE 10.2 YouTube

Image Type	Identification Accuracy Rate (%)	Verification	
		Error (%)	Hit (%)
Random	85.62	14.38	5.63
Eigen	91.43	8.57	7.45
Feature trained	93.12	6.88	9.11
Normal feature	90.67	9.33	3.29

TABLE 10.3 Facebook

Image Type	Identification Accuracy Rate (%)	Verification	
		Error (%)	Hit (%)
Random	85.10	14.9	1.39
Eigen	93.42	6.58	3.74
Feature trained	95.89	4.11	5.01
Normal feature	87.34	12.57	0.65

TABLE 10.4 Instagram

Image Type	Identification Accuracy Rate (%)	Verification	
		Error (%)	Hit (%)
Random	88.39	11.61	6.87
Eigen	98.54	1.46	2.49
Feature trained	99.11	0.86	9.35
Normal feature	90.67	9.33	4.34

KEYWORDS

- **algorithms**
- **artificial intelligence**
- **authentication**
- **image authentication**
- **preliminary score approach**
- **security enhancement**
- **self-database images**

REFERENCES

1. Ranjan, R., Bansal, A., Zheng, J., Xu, H., Gleason, J., Lu, B., Nanduri, A., et al., (2015). A fast and accurate system for face detection, identification, and verification. *Journal of Latex Class Files, 14*(8), 1–16.
2. Phillips, P. J., Grother, P., Micheals, R. J., Blackburn, D. M., Tabassi, E., & Bone, M., (2003). *Face Recognition Vendor Test 2002: Evaluation Report*. Tech. Report NISTIR 6965. NIST, 188-103.
3. Li, S. Z., & Jain, A. K., (2005). *Handbook of Face Recognition*. Springer-Verlag, Secaucus, New Jersey.
4. Li, H., & LiJing, J., (2010). Implement of face recognition system based on hidden Markov model. In: *IEEE, Sixth International Conference on Natural Computation (ICNC)*. 978-1- 4244.
5. Jen-Tzung, C., & Chih-Pin, L., (2008). Maximum confidence hidden Markov modeling for face recognition. *IEEE Trans. on Pattern and Intelligence*, (4), 606–616.
6. Rogier, N., & Chauvaud, P., (1997). MD2 is not secure without the checksum byte. *Designs Codes Cryptograph, 12*(3), 245–251.

7. Matsuo, T., & Kaoru, K., (2004). On parallel hash functions based on block-ciphers. In: *Proceedings of the IEICE Transactions on Fundamentals of Electronics, Communications and Computer Sciences* (pp. 67–74).
8. Xiaotie, D., Chan, L., & Huafei, Z., (1999). A proposal for secure hash algorithm. In: *Proceedings of the 1999 International Workshop on Cryptographic Techniques and E-commerce* (pp. 254–258).
9. Hampel, H., Arps, R., & Chamzas, C., (1992). Technical features of the JBIG standard for progressive bi-level image compression. *Signal Process Image Commun., 4*(2), 103–111.
10. Venkatesan, R., Koon, S., & Jakubowski, M., (2000). Robust image hashing. In: *Proceedings of the IEEE International Conference on Image Processing, 3*, 664–666.
11. Goljan, M., Fridrich, J. J., & Du, R., (2002). Distortion-free data embedding for images. In: *Proceedings of the 4th International Workshop on Information Hiding* (pp. 27–41).
12. Marvel, L. M., & Hartwig, G. W., (2000). Compression compatible fragile and semi-fragile tamper detection. In: *Proceedings of the SPIE International Conference on Security and Watermarking of Multimedia Contents* (Vol. 2, pp. 3971–3981).
13. Jang, D., Yoo, C. D., & Lee, S., (2005). An SVD-based watermarking method for image content authentication with improved security. In: *Proceedings of the IEEE International Conference on Acoustics, Speech, and Signal Processing* (Vol. 2, pp. 525–528).
14. Krawczyk, H., Gennaro, R., & Jarecki, S., (1996). Robust threshold DSS signatures. Advances in cryptology—EUROCRYPT. In: *Proceedings on the International Conference on the Theory and Application of Cryptographic Techniques; Lecture Notes in Computer Science* (Vol. 1070, pp. 354–371). Springer.
15. Tian, J., (2002). Reversible watermarking by difference expansion. In: *Proceedings of Workshop on Multimedia and Security* (pp. 19–22).
16. Fridrich, J., Goljan, M., & Du, R., (2001). Invertible authentication. In: *Proceedings of the SPIE, Security and Watermarking of Multimedia Contents* (pp. 23–26). San Jose, California.
17. Pfitzmann, B., (1996). Digital signature schemes: General framework and fail-stop signatures. *Lecture Notes in Computer Science, 1100.* Springer, Berlin.
18. Brassard, G., (1993). *Cryptologie Contemporaine*. Masson, Paris.
19. Matsuo, T., & Kaoru, K., (2004). On parallel hash functions based on block-ciphers. In: *Proceedings of the IEICE Transactions on Fundamentals of Electronics, Communications, and Computer Sciences* (pp. 67–74).
20. Skala, V., & Kucha, M., (2001). The hash function and the principle of duality. In: *Proceedings of the Computer Graphics International* (Vol. 200, pp. 167–174).
21. Trappe, W., & Washington, L. C., (2002). *Introduction to Cryptography: With Coding Theory*. Prentice-Hall, Upper Saddle River, New Jersey.
22. Goswami, G., Bhardwaj, R., Singh, R., & Vatsa, M., (2014). MDLFace: Memorability augmented deep learning for video face recognition. In: *IEEE International Joint Conference on Biometrics*.
23. Barr, J. R., Bowyer, K. W., Flynn, P. J., & Biswas, S., (2012). Face recognition from video: A review. *International Journal of Pattern Recognition and Artificial Intelligence* (Vol. 26, No. 05).

24. Mintzer, F., Braudaway, G., & Yeung, M., (1998). Effective and ineffective digital watermarks. In: *Proceedings of the ICIP'97*. Santa Barbara, CA, USA.
25. Yeung, M. M., (1998). Digital watermarking introduction. *Commun. ACM, 41*(7), 31–33.
26. Bhatt, H. S., Singh, R., & Vatsa, M., (2014). On recognizing faces in videos using clustering-based re-ranking and fusion. *IEEE Transactions on Information Forensics and Security, 9*(7), 1056–1068.
27. Kirby, M., & Sirovich, L., (1990). Application of the Karhunen- loève procedure for the characterization of human faces. *IEEE Trans. Pattern Analysis and Machine Intelligence, 12*, 831–835.
28. Darrell, T., Azarbayejani, A., Wrem, C., & Pentland, A., (1995). Real-time tracking of the human body. *IEEE Trans. on Pattern Analysis and Machine Intelligence, 19*, 780–785.
29. Chen, M., He, Y., & Kundu, A., (1992). Handwritten word recognition using hidden Markov models with adaptive length Viterbi algorithm. *IEEE Int. Conf. on Acoustics, Speech, and Signal Processing, 3,* 153–156.
30. Ben, A. J., & Nandy, D., (1998). A volumetric/ iconic frequency domain representation for objects with application for pose invariant face recognition. *IEEE Trans. on Pattern Analysis and Machine Intelligence, 20*, 449–457.
31. Muhammad, S., Jamal, H. S., Sajjad, M., & Mudassar, R., (2013). Sub-holistic hidden Markov model for face recognition. *Research Journal of Recent Sciences, 2*(5), 10–14.
32. Othman, H., & Aboulnasr, T., (2003). A separable low complexity 2D HMM with application to face recognition. *IEEE Trans. on Pattern Analysis and Machine Intelligence, 25*(10).
33. Bicego, M., Castellani, U., & Murino, V., (2003). Using hidden Markov models and wavelets for face recognition. *Proc. of IEEE International Conference on Image Analysis and Processing*, 52–56.
34. Portera, G., & Doran, G., (2000). An anatomical and photographic technique for forensic facial identification. *Forensic Science International, 114*, 97–105.

CHAPTER 11

MACHINE LEARNING AND CYBER DEFENSE: HYPE VS. REALITY

RENU, MEENA SACHDEVA, and INDERPREET KAUR

Department of CSE, GCET, Greater Noida, Uttar Pradesh, India

ABSTRACT

Modern businesses need to secure crucial assets, ranging from critical infrastructures to personal information, from cyber-attacks while achieving optimized business processes. Artificial intelligence (AI) is being used with huge potential to support information security. Machine learning (ML) can be utilized to prevent our information from a number of attacks, but this is not an acceptable solution. The technology also helps to combine all customer data, which serves as input to produce new vulnerabilities. Different opinions consider it as a game between attackers and security providers. In this chapter, we discuss the application areas, true potential, and limitations of ML. ML-based technologies are improving business processes by providing opportunities to design defense mechanisms against large-scale frauds. However, attackers can also misuse it to launch new threats. We demonstrate the ways of utilizing ML by both defenders and attackers. An extensive review and experiments performed on real enterprise systems are taken into account to draw conclusions.

11.1 INTRODUCTION

Machine learning (ML) works as an important tool in enhancing the protection measures in attack detection systems. ML algorithms are grouped

The Fusion of Artificial Intelligence and Soft Computing Techniques for Cybersecurity.
M. A. Jabbar, Sanju Tiwari, Subhendu Kumar Pani, & Stephen Huang (Eds.)

into supervised and unsupervised learning algorithms based on the input. Supervised learning relies on labeled training data, while unsupervised learning relies on unlabeled training data to infer the classes on its own [1]. A Comparative Survey covers the Influence of ML Techniques on cybersecurity. Various researchers and organizations have claimed that AI is not immune to attacks [2]. The chapter is divided into several sections. Section 11.2 illustrates the paradigms of ML algorithms for cyber defense. Section 11.3 outlines the application of ML algorithms for cybercrime. Section 11.4 includes the Survey and Evaluation, and Section 11.5 covers the current trends of ML algorithms in the cyber industry. Finally, Section 11.6 concludes the chapter with user awareness.

11.2 PARADIGMS OF MACHINE LEARNING (ML) AND APPLICATIONS IN SECURITY

ML algorithms are commonly referred to as shallow learning (SL) and deep learning (DL). Both categories are utilized in cybersecurity, as shown in Figure 11.1.

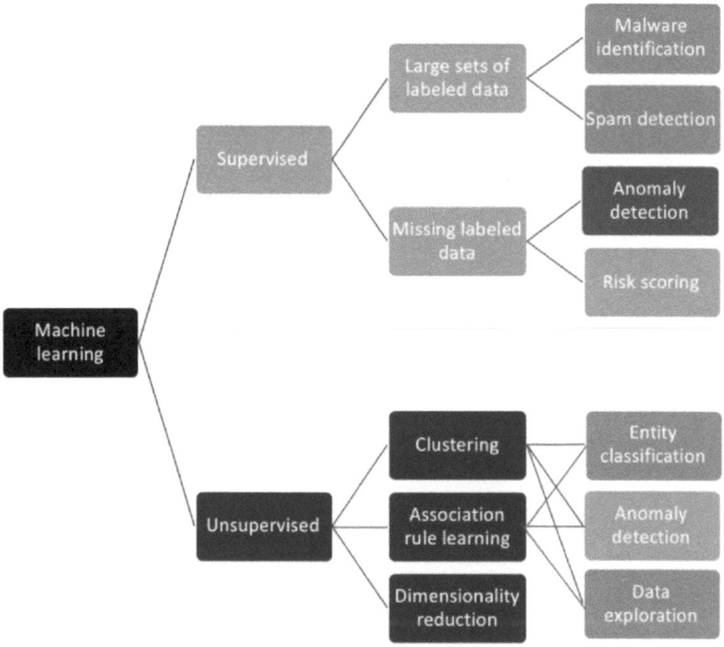

FIGURE 11.1 Machine learning algorithms and applications in cyber security.

Shallow techniques require a human expert who can identify the relevant data and its characteristics before the final implementation of the algorithm. On the other hand, DL works with multi-layered input data and automatically selects features through unlabeled learning [3]. Both SL and DL encompass numerous techniques. Popular techniques for SL algorithms include K-nearest neighbor (KNN), support vector machines (SVMs), naïve bayes (NB), logistic regression (LR), random forest (RF), hidden Markov models (HMMs), and shallow neural network (SNN) [4]. These techniques are mostly used for classification and are sometimes combined by a human expert. Deep neural networks (DNNs) serve as the foundation for any DL algorithms and consist of large neural networks capable of autonomous learning. Neurons in one layer may be fully connected, forming fully-connected feedforward deep neural networks (FNN), or partially connected to only a subset of the next layer, forming convolutional feedforward deep neural networks (CNN).

11.3 MACHINE LEARNING (ML) FOR CYBER CRIME

Criminals are also taking advantage of this tool in crafting malicious activities and targeting thousands of people every day. Performing such activity manually is a tedious and time-consuming job; hackers can automate this with the help of ML. These activities are categorized into 5 groups based on their target and level of harm:

1. **Information Gathering:** This is the start of every cyberattack. More information means a higher probability of success in a cyberattack. Collected information can be organized and gathered both online and offline. Neural networks and their associated clustering and classification methods can be used to analyze victims' posts on social networks. Modern networks are software-defined networks (SDN), where ML can assist adversaries. Hackers can gather information about security tools and other quality of service (QoS) related policies. An attacker can benefit from analyzing traffic and flow rules. Recently, the Social Mapper tool was designed to locate individuals across various social platforms using Google images.
2. **Impersonation:** It allows attackers to target victims through various methods such as spamming, phishing, and spoofing. One of the oldest and most popular forms is email spam, where ML has

been employed by cyber attackers. By training a neural network, the process of spamming can be automated, generating junk emails without raising suspicion. Impersonation through email is relatively simple, and a person's behavior can be easily tracked through their posts. Recently, the Markov model has been utilized in conjunction with recurrent neural networks (RNNs) to generate new tweets based on previous ones.

3. **Unauthorized Access:** To gain unauthorized access is common in two popular areas where ML can be of assistance. Captcha bypassing is the most popular method in mass hacking, where object segmentation is achieved through the support vector machines (SVMs) method. ML is utilized to break reCAPTCHA images with an accuracy of 82% [7]. Another area is password brute force, where passwords can be guessed with the help of ML. Markov models were first trained to generate a pattern with a set of similar passwords.
4. **ML for Attack:** Most of the attacks are performed due to phishing, which leads to the download of malware, spyware, ransomware, or other malicious programs. Fuzzing is a common practice among hackers to discover system vulnerabilities. With the help of AI-based smarter techniques, new examples of documents can be rapidly found to crash any application. Additionally, vulnerability discovery can be accelerated in a cost-effective manner by using ML-based algorithms.
5. **Automate the Crime:** Criminals can automate tasks involved in various activities. While there is no specific prediction about what will be automated, organizations related to cybercrime are hiring experts to develop software for their support.

11.4 LITERATURE SURVEY AND EVALUATION

This section will cover the benefits and side effects of ML. It will also provide guidelines for industries to decide whether to apply ML algorithms. Currently, no algorithm can be applied without human intervention. There are two aspects that make ML a good choice for the industry. First, a large volume of useful data can be easily collected to provide an initial level of security. Various tools are designed to help sort, slice, and mine large volumes of data automatically, assisting security analysts. A lot of work has been done on intrusion detection systems (IDS) using ML. Understanding the data source is the first step for any security policy. Various methods have been tried to apply ML to IDS by different researchers. A very popular way of attack detection is Packet-Based, where packets provide all the communication-related details. Packet information is parsed first with the help of unsupervised learning. A fuzzy C-means based algorithm was introduced by Hu et al. [8]. Yang Tian-Yi et al. [9] proposed packet parsing methods and payload analysis methods. Aaron Adler et al. proposed Behavior-Based Access Control with attack Detection on TCP Connections and HTTP Requests. Gong et al. [10] provided a Feasibility Analysis of Data Transmission in SDN. Another problem is the availability of qualified, experienced security analysts. Designing efficient defensive policies is a complex game, where a single wrong decision is enough to open the door for vulnerabilities. AI-driven attacks are becoming very frequent, creating a demand for security experts in the current world, which is full of data everywhere. In the cyber world, organizations are still hesitant to trust ML-based defensive mechanisms. ML has been widely used by both hackers and security communities [11]. Naturally, trust with your peers is preferred compared to trust with a ML model. In the upcoming years, a combination of ML models and security experts will be preferred.

11.5 CURRENT TRENDS OF MACHINE LEARNING (ML) ALGORITHMS IN CYBER SECURITY

Currently, a large majority of various ML approaches are being utilized to design "warning" systems, but they still require human experts for the final decision. Intrusion detection, malware analysis, and spam detection are three major areas where ML algorithms are applied [12]. ML models are being used as a defense under different names, some of which are popular and listed here:

1. **Splunk:** It is a software that has a range of applications, from simple IT operations to analytics and cybersecurity. ML is used in User Behavior Analytics to identify digital weak points of users and respond to malware attacks quickly and automatically.
2. **Chronicle:** It is a software that has a range of applications, from simple IT operations to analytics and cybersecurity. ML is used in

User Behavior Analytics to identify digital weak points of users and respond to malware attacks quickly and automatically.

3. **Microsoft:** Windows defender, named "advanced threat protection" (ATP), is exclusively designed for prevention, breach detection, and automated recovery. It utilizes cloud AI and ML algorithms at multiple levels to identify and mitigate threats.

11.6 CONCLUSION

Cyber attackers may target us anywhere, in any form. Important products or tools can be hijacked for the sake of money, or criminals can access valuable data that may go unnoticed for months. Machine learning (ML) is helping in the cyber world by providing security solutions and solving industry-specific problems, but not every problem. On the opposite side, criminal activities are also benefiting from ML to perform cybercrime faster and cheaper. A critical challenge in cybersecurity is to design adaptable and automated security policies to target new threats and vulnerabilities more effectively. It is very important to decide when and which algorithms can provide mature enough solutions. This chapter may help future researchers analyze existing solutions and the various issues of ML in sensitive environments. A strong security plan can be implemented in multiple layers of protection throughout computers, programs, and networks.

KEYWORDS

- **advanced threat protection**
- **algorithms**
- **cyber attackers**
- **cyber defense**
- **cybersecurity**
- **deep learning**
- **machine learning**
- **security**

REFERENCES

1. Ghasem, Z., Frommholz, I., & Maple, C., (2015). Machine learning solutions for controlling cyberbullying and cyberstalking. *J. Inf. Secur. Res.*, *6*(2), 55–64.
2. Mbaziira, A., & Jones, J., (2016). A text-based deception detection model for cybercrime. In: *Int. Conf. Technol. Manag.*
3. Sultana, N., Chilamkurti, N., Peng, W., & Alhadad, R., (2019). Survey on SDN-based network intrusion detection system using machine learning approaches. *Peer-to-Peer Networking and Applications*, *12*(2), 493–501.
4. Prabakaran, S., & Mitra, S., (2018). Survey of analysis of crime detection techniques using data mining and machine learning. In: *Journal of Physics: Conference Series* (Vol. 1000, No. 1, p. 012046). IOP Publishing.
5. Ford, V., & Siraj, A., (2014). Applications of machine learning in cyber security. In: *Proceedings of the 27th International Conference on Computer Applications in Industry and Engineering.*
6. Yavanoglu, O., & Aydos, M., (2017). A review on cyber security datasets for machine learning algorithms. In: *2017 IEEE International Conference on Big Data (Big Data)* (pp. 2186–2193). IEEE.
7. Apruzzese, G., Colajanni, M., Ferretti, L., Guido, A., & Marchetti, M., (2018). On the effectiveness of machine and deep learning for cyber security. In: *2018 10th International Conference on Cyber Conflict (CyCon)* (pp. 371–390). IEEE.
8. Zen. (2021). AI and machine learning in cyber security. In: *Towards Data Science: A Medium Publication Sharing Concepts, Ideas, and Codes.*
9. Tian-Yi, Y., et al., (2019). Network traffic anomaly detection based on incremental possibilistic clustering algorithm. *J. Phys.: Conf. Ser.*
10. Michael, J. M., Michael, A., Aaron, A., & Rachel, G., (2015). *Use of Machine Learning in Big Data Analytics for Insider Threat Detection, 10*(1), 120.
11. Aaron, A., Michael, J. M., Jeffrey, C., Michael, A., & Rachel, G., (2013). Using machine learning for behavior-based access control: Scalable anomaly detection on TCP connections and HTTP requests. In: *32nd Military Communications Conference (MILCOM 2013).* San Diego, CA.
12. Gong, S., Siddiqui, M. K., Luo, Y., et al., (2017). Feasibility analysis of data transmission in SDN. *Journal of Intelligent and Fuzzy Systems, 33*(5), 3145–3152.

CHAPTER 12

AN EFFECTUAL INVESTIGATION ON SECURITY AND PRIVACY IN VEHICULAR AD-HOC NETWORKS WITH PREVENTIVE MEASURES

SHALLY NAGPAL, SHIVANI GABA, ALANKRITA AGGARWAL, and DEEPAK KUMAR

Panipat Institute of Engineering and Technology, Panipat, Samalkha, Haryana, India

ABSTRACT

Vehicular Ad-hoc Networks (VANETs) can make streets more secure, cleaner, and more brilliant. It can offer a wide extend of administrations, which can be security and non-safety related. Numerous safety-related VANETs applications are real-time and mission basic, which would require strict ensure of security and unwavering quality. Indeed non-safety related interactive media applications, which is able play an critical part within the future, will require security back. Need of such security and protection in VANETs is one of the key preventions to the wide spread usage of it. An unreliable and untrustworthy VANET can be more dangerous than the framework without VANET bolster. So it is fundamental to create beyond any doubt that "life-critical safety" data is secure sufficient to rely on. Securing the VANETs together with suitable assurance of the security drivers or vehicle proprietors may be a exceptionally challenging errand. In this chapter we summarize the assaults, comparing security necessities and challenges in VANETs.

The Fusion of Artificial Intelligence and Soft Computing Techniques for Cybersecurity.
M. A. Jabbar, Sanju Tiwari, Subhendu Kumar Pani, & Stephen Huang (Eds.)

12.1 INTRODUCTION

Vehicular ad hoc networks (VANETs) receive significant attention from academics and the automobile industry. VANETs present an opportunity to enhance road safety by facilitating communication between vehicles, without the need for additional fixed infrastructure. VANET applications can be categorized into three subcategories: safety applications, non-safety applications, and basic applications. Safety applications focus on protecting users and informing them about road incidents such as traffic collisions. Non-safety applications aim to improve user convenience and contribute to traffic optimization. Furthermore, VANET provides real-time road safety information to assist drivers and road users in staying safe. However, the VANET ecosystem is not immune to threats, as it is susceptible to vulnerabilities inherent in vehicular networks.

Dedicated short-range communications (DSRC) operating in the 16.9 GHz frequency band is allowed with the wireless access in vehicular environment (WAVE) protocol, which is built on the IEEE 802.11p standard. Most vehicle-to-RSU connectivity is achieved using wireless networking, also known as vehicular wireless access or WAVE-based devices. WAVE connectivity ensures that travelers' vehicles have up-to-date relevant information and accurately tracks the movement of traffic, improving pedestrian and rider safety. Additionally, it enhances traffic flow and the performance of the traffic management system. In the VANET, an OBU, an RSU, and a TA are units responsible for collecting valuable vehicle details such as speed, acceleration, power, and more [1]. The network then transfers this information to local vehicles through a wireless connection. RSU units are also connected to the TA through a wired network. The TA serves as the chief of all elements and is responsible for maintaining the VANET's proper functioning.

Figure 12.1 depicts a classical scenario of a wireless-based vehicular network in which various vehicles are interconnected by wireless technologies [2].

1. **Remote Specific User (RSUs):** Computer machines fixed alongside the highway or in designated locations such as parking lots or intersections. Cables are commonly used to help vehicles traveling on nearby roads stay connected. The RSUs include network equipment using IEEE 802.11p radio technologies for dedicated short-range networking (DSRC). Furthermore, RSUs are a particularly effective method of communication with other computers in other networks [3].

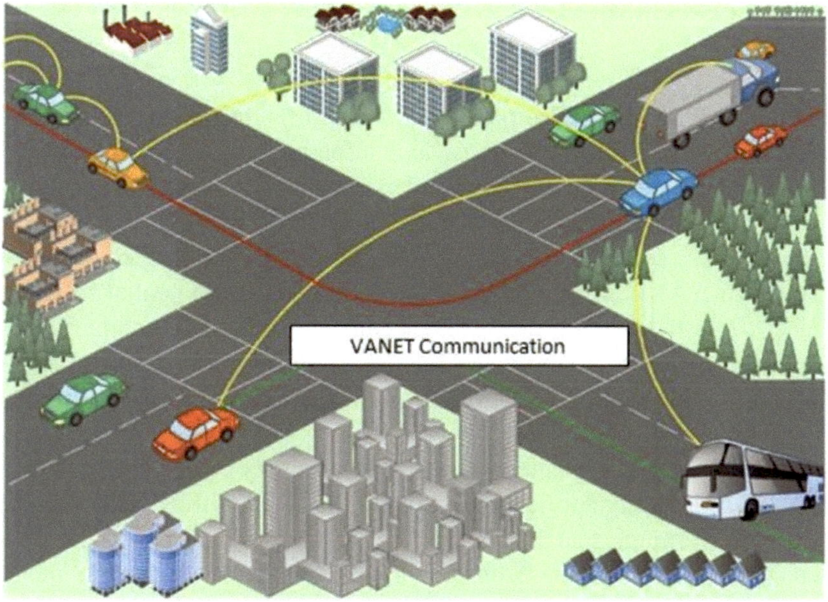

FIGURE 12.1 Scenario of wireless-based vehicular network.

2. **Onboard Unit (OB):** A GPS-based tracking system is almost always installed in any vehicle, allowing RSUs and other OBUs to monitor shared vehicle details. The RCP, sensors, interface, and read/write storage are all part of the electronics package that make up the OBU. The OBU has the primary responsibility to connect to another OBU through an IEEE 802.11p wireless connection. Additionally, the car battery serves as the primary power source for the OBU, and each vehicle contains sensors to monitor location and environmental conditions, an event data recorder (EDR), and motion sensors, all of which help provide feedback to the OBU [4].
3. **Trusted Authority (TA):** It is responsible for the entire VANET scheme, including registering the network resources, providing operational guidelines, and overseeing vehicle use. It is also required to verify vehicle authentication, user ID, and OBU ID to ensure VANET protection and prevent any vehicle damage. In the event of any malicious messages or unusual activity, the TA will expose OBU ID and information [5]. The TA system is also capable of identifying the criminals.

12.2 PERFORMANCE, PRIVACY, AND SECURITY ISSUES IN VANET

Because of the diverse implementations and facilities, such as improved traffic quality, passenger safety, and infotainment, VANETs have been extensively studied. Vehicle-to-everything (V2X) technologies must therefore address numerous technological challenges as they are less flexible, less capable, and provide low bandwidth and limited computing power. In VANETs, cloud computing is considered a potential solution to meet these requirements. However, conventional cloud computing may not meet the next-generation VANET standards for autonomous vehicles, which require high mobility, low latency, real-time implementations, and networking [6].

Most VANET implementations are either safety or non-safety oriented. In VANETs, safety applications are used to provide safety-related signals, such as warnings that help road vehicles respond to avoid dangerous conditions and incidents. Safety information includes road traffic updates, traffic jams, maintenance work, and emergency vehicle warnings. Low latency and high reliability are essential for these safety applications. On the other hand, non-safety uses of VANETs are very convenient and effective. Traffic control applications can be divided into two categories: protection and entertainment. In traffic management, software is used to optimize traffic flow and reduce congestion on the route. Informational and entertaining technologies are used to provide internet connectivity for travelers, such as storing records, downloading videos, and making phone calls. However, disseminating vital incident records, such as accident reports, accurately and rapidly is crucial in VANETs. Although VANETs can be used to disseminate event information, it is challenging to deliver timely and accurate information to a specific location while vehicles are in motion. This is due to the relatively small transmission range of DSRC and the contention-based carrier-sense multiple access scheme in IEEE 802.11p. MAC contention causes delays in vital messages, which is not ideal for VANETs. Failure to disseminate time-sensitive information in a timely and precise manner may result in collateral damage to nearby vehicles. The feasibility of cellular networks to support VANET services while providing large coverage and high data rates to vehicular users has recently been examined [7] (Figure 12.2).

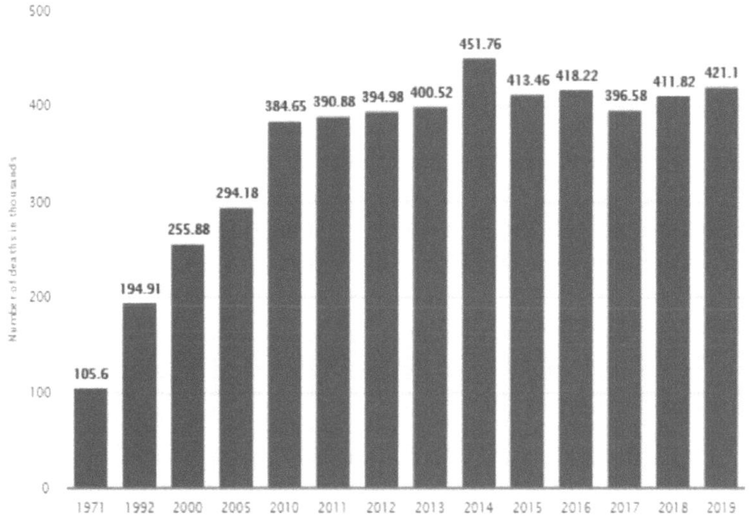

FIGURE 12.2 Graph of number of deaths in respective years.

In contrast, while a single wireless communications technology, such as DSRC or a cellular network, can support some ITS services in an area with heavy loads, high mobility, and a complex network topology, a complete ITS support from a single wireless communications technology, such as DSRC or a cellular network, might not be possible. A heterogeneous vehicular ad hoc network can be a potential medium that can satisfy many challenging connectivity criteria by combining cellular networks and IEEE 802.11p. Seamless data communication can be established between vehicles that are spatially separated in VANETs [1].

Traditionally, VANET technology has not been adequate for processing smart vehicle data including video and sensor data, which can produce enormous volumes of traffic data. Additional servers are needed in distributed areas in order to capture and process a vast volume of instantaneous traffic information. The cloud computing approach, coupled with VANETs, could be the best option for this kind of use.

Recent studies show that mobile users are now making use of cloud computing for complex computations that were difficult or impossible to perform locally. There have been a number of studies conducted on VANETs that incorporate cloud computing in order to advance the potential of VANETs even further. Most of the vehicles' uploads and downloads use time and resources [8]. Currently, cloud storage cannot meet the location recognition, accessibility assistance, and latency standards for vehicle density in urban areas.

12.3 CHALLENGES IN VANET-BASED ENVIRONMENT

Due to vulnerabilities in VANET that cause information communication vulnerabilities, VANET is susceptible to attacks. In order to ensure vehicle-to-vehicle (V2V) and network stability, VANET security systems must meet specific requirements. While all networks must adhere to the same standards, only certain networks are required to do so for VANET. The conditions are as follows:

1. **Authentication:** VANET should require vehicles to authenticate each other in order to enable V2V contact. This method can involve identifying the sender's identity and verifying the authenticity of the sender.
2. **Availability:** In this context, availability refers to the VANET system's ability to be functional and operational at all times. Like any system, response time is an important consideration for any program.
3. **Privacy:** It is critical in VANET. The name and location of the vehicle must be kept secret.
4. **Integrity:** It refers to trust. In a free flow of content, there can be no modification attacks. Knowledge is reliable because of this.

A powerful, clean, insightful, and entertaining mass transit system relies on VANETs, which are also one of the foundational technologies in the field. Today, the time people spend in cars is huge. Smart vehicles based on VANETs have emerged to make the time more stable, reliable, friendly, pollution-free, and economical. More protection can be created by ensuring that important incidents, such as those related to safety, are reported accurately. Efficient transport is made possible by reducing road congestion and noise, and by improving travel time predictability. When you take VANETs on the go, they will connect to the Internet and provide you with downloadable information, as well as links to social networks.

The signals VANETs use are either beacons or protection messages. Vehicles use transmitters to transmit and relay status information at 100-millisecond intervals. Beacons are sent to nearby vehicles and include the velocity, position, and pseudo-ID of the sender. Another perspective is that safety alerts often serve as emergency information for drivers on the road, helping them avoid potentially life-threatening conditions and enabling timely rescue if needed. Vessels broadcast a WAVE short message (WSM) as they encounter events on the route. The message payload includes vehicle position, message sending time, vehicle location, and local road activities. In vehicular contact range, each vehicle collects information about the other vehicles. We consider heterogeneous VANETs, where vehicles use IEEE802.11p and LTE cellular technology, to be heterogeneous as well [9].

Virtual LANs (VLANs) use remote specific user (RSUs) as gateways to connect to cloud providers on the go. Traffic and multimedia data are acquired by vehicular nodes through cloud services. Vehicles are located on the outer edge of the network, while the cloud is at the center. With a cloud-based VANET, it is possible to track the location of all cars in the network and estimate traffic patterns in real-time, enabling personalized safety alerts for those in need. While VANETs leverage the internet to address current challenges, the expansion of smart vehicles still requires addressing certain fundamental issues. Future VANETs are defined as a result of the rapid growth and improvement of VANETs [10].

12.4 ASSAULTS AND THREAT PERSPECTIVES

Due to inherent characteristics such as high mobility and constant disconnection, VANETs are more vulnerable to attacks. The mobility of nodes in cars and the star topology of VANETs vary rapidly. While nodes in MANET simulations move at a typical speed of 20 m/s, vehicle speeds in VANETs are even higher. This also means that V2V network interruptions are common. In particular, cars going in different directions are only able to interact with each other for a few seconds, and these gaps are regularly filled with a lack of connection in the vehicular network. Since the cars don't see each other again, they don't know who they are racing towards. This makes VANETs easier to target and harder to identify as suspicious cars [11] (Figure 12.3).

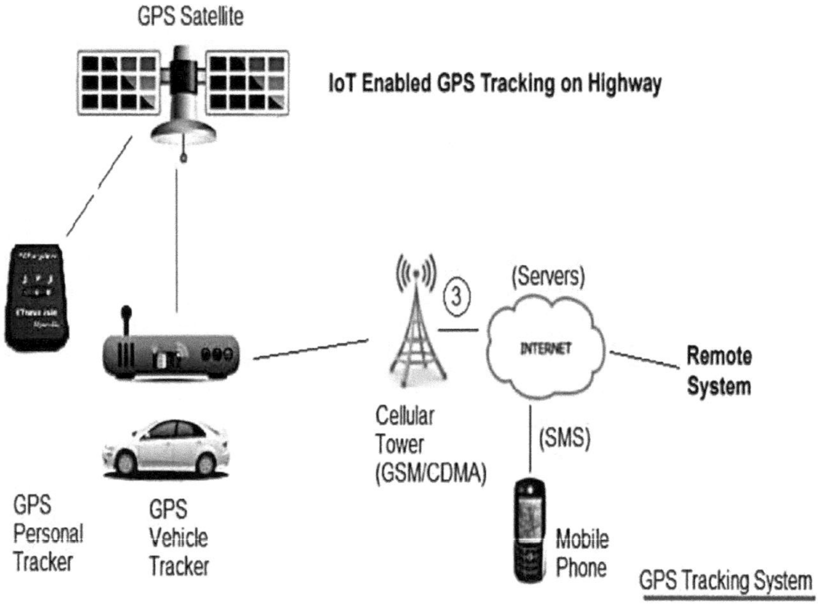

FIGURE 12.3 GPS-based advanced internet of vehicles.

To ensure the dissemination of safety warnings in an open access framework, the integrity of contacts in VANETs is disrupted by attacks that involve message modification, injection, or interception. These attacks pose a significant risk to the security of VANETs, as attackers can manipulate the information in messages to further exploit vulnerabilities. VANETs face various types of threats and challenges in dealing with these dangers [12].

In VANETs, intruders aim to harm the entire network for their own gain. There are four distinct classifications of attackers in VANETs. An intruder is an attacker who is authenticated within the network, while an outsider is a non-authenticated attacker who lacks the capabilities to access the network. Active attackers are those who actively send or generate false messages, while inactive attackers do not transmit or deliver messages. Additionally, attackers can be categorized as malicious or rational. Malicious attackers seek to cause maximum damage without any personal benefits, whereas rational attackers aim to gain benefits. Local attackers employ a small number of vehicles to carry out attacks, while expanded attackers utilize a complete set of vehicles to target multiple networks.

It is crucial for VANETs to function securely, as a substantial understanding of threats and attacks is required. In this context, we will discuss the risks and attacks on each of the different security services found in VANETs [13].

12.4.1 LACK OF ACCESS

Accessibility of information is critical in the VANET method. In the event of accessibility issues, the reliability of VANETs can be compromised. This segment analyzes the threats and attacks on the availability of VANETs.

Denial of Service (DoS) attacks include vehicle-based denial of service attacks and external attacks on the network vehicles. Jamming of V2V communication can result in a roadblock of all available options in the networks. Distributed denial of service (DDoS) is also known as many-to-many DDoS.

12.4.2 JAMMING ATTACK

In this scenario, the intruder utilizes a highly powerful signal that operates on the same frequency as the existing network transmission link. The main vulnerability of these devices lies in the lack of a safety signal, as the threat remains undetectable. For a jamming attack to be successful, the jammer must effectively execute an operation that simultaneously blocks the valuable signal [14].

12.4.3 MALICIOUS SOFTWARE INTRUSION

The machine can be attacked through the program that supports the OBU and RSU devices. In the event of a malware attack in VANET, it is possible that the other features of the VANET framework will become inoperable.

12.4.4 BROADCAST TAMPERING ATTACK

Untrustworthy vehicles may manipulate the same messages by altering the message information or creating and inserting a message within the VANETs. As a result, it is possible for hazardous incidents to remain undetected, which could be a contributing factor to accidents.

This is the most commonly used attack in the room, and it appears in both ad hoc and VANET environments. The most common form of attack on VANETs is carried out by registered users of VANETs. In order to determine whether or not the accused node is functioning correctly, the packets are sent to it, and it fails to accept the networking contribution. The malicious node will interfere with the routing table, and this would prevent an important message from being sent to the recipients according to the node pretending to help with the non-practical case [15].

The form of active attacks includes malicious cars, such as Smurfs and spoof packets. These Smurfs collect sensitive data, but instead of forwarding it, they dump it. Malicious vehicles abuse the MAC protocol to expand bandwidth and cost other users, known as a Greedy Behavior Attack. As a result, traffic on the transmitting channel overflows, causing a crash that disrupts the legitimate facilities of authorized customers. An attacker floods the VANET with a huge volume of spam messages, such as commercials, overloading the network by consuming more bandwidth.

Confidentiality can be assured by using credentials and distributing public keys on all exchange messages. This makes it impossible for vehicles outside of the nodes to comprehend private and classified information being communicated among the vehicles. A secure cryptographic solution protects confidentiality. Throughout this segment, we will cover the most commonly seen confidentiality risks [16].

Wired networking systems such as MANETs and VANETs often face eavesdropping attacks. The main objective of this attack is to access secure data with sensitive information. Due to this attack, information that is private to registered individuals, such as identification and location data, can be accessed by unregistered users and used to track vehicles. It is one of the classified threats that pose a challenge to traffic analysis. During this attack, the attacker analyzes the frequency of message transmission and collects as much information as possible.

12.4.5 MAN-IN-THE-MIDDLE ATTACK

In some cases, V2V contact occurs in the middle of a man-in-the-middle attack to monitor and potentially change the communications. Access and power are gained by the intruder, but contact institutions believe they should engage with each other privately.

12.4.6 SOCIAL DISTRACTION

It is used to divert the driver's attention from the road. When it comes to drivers, the perpetrator has an in-depth understanding of immorality and dishonest conduct. An attacker's goal is to elicit responses from vehicles after sending messages of this nature. As a result, it affects the driving and efficiency of vehicles in the VANET scheme. A critical aspect of the VANET protocol is authentication, which is used to combat malicious nodes that may attempt to infiltrate the system. VANETs are protected from both internal and external attacks through authentication. In this section, we will explore the authentication risks in VANETs.

The first reference to the Sybil Attack is mentioned in a previous study. In this scenario, a node sends out multiple messages using false IDs. By employing certain techniques, the intruder can interfere with the vehicles' operating systems and then utilize a V2V message mechanism to collect their responses. As a result, congestion is created on the track, leading to the need for altered routes in order to avoid obstructing other vehicles. Close to the wormhole invasion, there is a known tunneling attack. The attacker sets up a private communication link by utilizing the same infrastructure to initiate the link and then connects two different points in VANETs through a tunnel. This allows nodes that are far apart to communicate with each other. In this attack, the intruder employs deceptive techniques to generate fake GPS coordinates, tricking vehicles into believing that the coordinates are valid in another area.

12.4.7 NODE IMPERSONATION ATTACK

When an attacker mistakes the correct ID of the unique visitors in a VANET, they are taking part in a Node Impersonation Attack. False authentication attempts are made when being connected with a cooperative secure communication. The malicious user can make use of other users' authentication contributions, and thus circumvent its own authentication in this attack. In this case, it is an example of free-riding action. A substantial challenge can be posed to cooperative digital signatures if this attack is effective.

A playback attack is one of the most often used attacks as well. This attack happens when legitimate data is distributed fraudulently or causes a pause to be created, resulting in an unwanted and malicious effect. In order to overcome this attack, the VANET requires a large amount of memory for a cache and a greater number of time sources to compare the messages.

Vehicle ID attacks are triggered by the use of duplicate codes and/or credentials of other cars for authentication purposes. The primary goal of this attack is to provide false information to trustworthy authorities (TAs) in order to generate doubt [17]. Attack methods include message tampering in V2V or V2I communication.

To disguise their movements, intruders use fake identities to assume the identity of another vehicle. This attack occurs when a user fails to display their own identity and instead impersonates another user to gain unauthorized access.

12.5 COMPROMISE IN DATA INTEGRITY

Ensuring data confidentiality during transfer is the only means to maintain the integrity of the data being transmitted. We will cover the various risks to data integrity in this section. A message that appears to originate from a registered node is being broadcast. In case of an emergency, identifying cars during the traffic jam may be problematic because the malicious perpetrator plans to repeat or pause the transmission by providing valid data and injecting beacons and messages received before a VANET network.

12.5.1 ERROR TAMPERING ASSAULT

As the intruder modifies or adjusts recent message data to be sent, this attack usually occurs. An attacker can, for instance, reroute traffic in areas with congestion. This could potentially impact users' behavior when it comes to their driving.

12.5.2 ILLUSION ATTACK

This attack gathers information from surrounding antennas and uses traffic light sensors to emit false traffic warning signals due to a hazard on the lane. Vehicle collisions and traffic congestion will result from the illusion attack. Additionally, it aims to reduce the efficiency of the VANET system by utilizing unfavorable bandwidth.

12.6 NON-REPUDIATION ASSAULT

This attack occurs when the perpetrator prevents all senders and receivers from denying that messages were sent and received. The perpetrator will not communicate in case of any conflict.

12.6.1 KEY PERSPECTIVES AND DIMENSIONS

Due to the significant potential in assisting motorists and passengers in maintaining their health while on the road, vehicle automation has garnered considerable attention from the ITS research community and has become one of the most prominent research areas in intelligent transportation systems (ITS).

The ITS is actively working towards providing reliable connectivity to enhance traffic management, road safety, and alleviate traffic congestion [18]. The ability to communicate V2X offers substantial benefits to the ITS, including improving traffic quality, reducing traffic fatalities, and providing real-time updates to enhance driving experiences, such as accident alerts, traffic slowdown information, traffic jam warnings, and other transportation services.

Transmission rate and latency are crucial characteristics of V2V communication. To promote V2V communication, vehicles can relay helpful information such as emergency braking, crash detection, and traffic conditions. Vehicle-to-Infrastructure (V2I) communication enables vehicles and network infrastructures to exchange valuable information.

This network connects with many other networks, including the Internet, to transmit information between vehicles and RES systems. Since they have a direct link to the networks, V2I systems need more bandwidth than V2V systems, but they are not susceptible to attack.

Now, V2X technology has recently become available. The V2X connectivity network aims to facilitate V2V communications. The 3GPP-created V2X is considered the networking technology that can perform C-V2X communications. It is capable of integrating all vehicles and, along with it, is responsible for the framework of cooperative intelligent transport systems (C-ITS) that help reduce traffic congestion and improve traffic quality.

LTE-V, long-term evolution (LTE), and C-V2X were specifications that were first announced in 2016, and they are now referred to as LTE-V, LTE, and C-V2X. The high data rate, broad coverage, and penetration rate of LTE support stable V2I communication because of this. Despite LTE's decentralization, it faces several obstacles in V2V communication due to the lack of sufficient resources to facilitate V2V communication.

When the message is requested to be sent to a single vehicle or to a group of vehicles, use this notification. As an example, an alert will be sent to the drivers still on their route to stop traffic delays and improve traffic safety if an accident or crash occurs. The method for sending the alert messages to just the destination is a new routing algorithm, which is needed to manage this problem [19].

12.6.2 VEHICLE-TO-VEHICLE (V2V) GROUP COMMUNICATION

Only vehicles that share common features can participate in this communication. This includes vehicles of the same brand and their respective locations at a given time. Vehicle Beaconing: Vehicles in the vicinity receive periodic signals, and RSUs are also notified. These messages contain information about the speed, direction, and speed of the receiving vehicle.

If there is a likelihood of collision or accident, warnings for vehicular traffic are transmitted from the connectivity via RSUs to all vehicles within the vicinity. This is especially important when the roadway is curving, an intersection is encountered, or the road is narrow.

12.7 CONCLUSION AND FUTURE SCOPE

One of the main elements of wireless architecture is VANET, in which vehicular devices are interconnected with advanced protocols and algorithms. In some cases, Intelligent Transportation Networks are used with VANETs for Intelligent Communications on the road. An understanding is given to the notion that they have developed into something more expansive: "the Internet of Vehicles (IoT)." This integration predicts that these larger concepts will eventually develop into a concept with the name "the Internet of autonomous vehicles." Wireless networking technologies, such as Wi-Fi, can be used as the basis for VANETs or Wi-Fi, medium-range radio technologies like AMI broadcasting, and long-range radio technologies like SAT-radio (either standard Wi-Fi or ZigBee). Cellular networks or LTE are also useful for VANETs. Visible light contact (VLC) is the latest in this networking technology (Infrared transmission and reception). The rapid use of vehicular ad hoc networks (VANETs) is bringing the internet of things (IoT) to the forefront. The manuscript presents the assorted perspectives of VANET with real-world assaults, which should be removed so that the cumulative environment will be safe and secure from different assaults. The future scope of the work is integrated with the usage of blockchain-based technologies and protocols with a security-based environment. Nowadays, blockchain technologies are quite prominent for a security-aware environment; the work can be associated with advanced implementations of blockchain.

KEYWORDS

- **blockchain**
- **internet of things**
- **malicious software intrusion**
- **node impersonation attack**
- **security-aware environment**
- **vehicular ad hoc networks**
- **vehicular devices**

REFERENCES

1. Kumar, A., Varadarajan, V., Kumar, A., Dadheech, P., Choudhary, S. S., Kumar, V. A., & Veluvolu, K. C., (2021). Black hole attack detection in vehicular ad-hoc network using secure AODV routing algorithm. *Microprocessors and Microsystems, 80*, 1033162.
2. Yahiabadi, S. R., Barekatain, B., & Raahemifar, K., (2019). TIHOO: An enhanced hybrid routing protocol in vehicular ad-hoc networks. *EURASIP Journal on Wireless Communications and Networking, 2019*(1), 1–19.

3. Akhter, A. F. M., Ahmed, M., Shah, A. F. M., Anwar, A., Kayes, A. S. M., & Zengin, A., (2021). A blockchain-based authentication protocol for cooperative vehicular ad hoc network. *Sensors, 21*(4), 1273.
4. Almalki, M. N. S., Challoob, A. L., & Al-shareeda, M. A., (2021). ID-PPA: Robust identity-based privacy-preserving authentication scheme for a vehicular ad-hoc network. In: *Advances in Cyber Security: Second International Conference, ACeS 2020* (Vol. 1347, p. 80). Penang, Malaysia, Revised Selected Papers. Springer Nature.
5. Rajeshwari, M. R., (2021). Privacy preserving data transmission in malicious vehicular ad-hoc Networks. *Turkish Journal of Computer and Mathematics Education (TURCOMAT), 12*(3), 167167–16761.
6. Ramamoorthy, R., & Thangavelu, M., (2021). Group-based dual mode key management scheme for secure communication in vehicular ad-hoc networks. *Wireless Personal Communications*, 1–216.
7. Sheikh, M. S., Liang, J., & Wang, W., (2020). Security and privacy in vehicular ad-hoc network and vehicle cloud computing: A survey. *Wireless Communications and Mobile Computing, 2020.* Statista Research Portal, URL Statista.com 2021.
8. Chaubey, N. K., & Yadav, D., (2020). A taxonomy of Sybil attacks in vehicular ad-hoc network (VANET). In: *IoT and Cloud Computing Advancements in Vehicular Ad-Hoc Networks* (pp. 174–190). IGI Global.
9. Manivannan, D., Moni, S. S., & Zeadally, S., (2020). Secure authentication and privacy-preserving techniques in vehicular ad-hoc networks (VANETs). *Vehicular Communications, 216*, 100247.
10. Al-Shareeda, M. A., Anbar, M., Hasbullah, I. H., Manickam, S., & Hanshi, S. M., (2020). Efficient conditional privacy preservation with mutual authentication in vehicular ad hoc networks. *IEEE Access, 8*, 1449167–144968.
11. Ali, I., Hassan, A., & Li, F., (2019). Authentication and privacy schemes for vehicular ad hoc networks (VANETs): A survey. *Vehicular Communications, 16*, 416–461.
12. Wu, L., Sun, Q., Wang, X., Wang, J., Yu, S., Zou, Y., & Zhu, Z., (2019). An efficient privacy-preserving mutual authentication scheme for secure V2V communication in vehicular ad-hoc network. *IEEE Access, 7*, 16160160–1616063.
13. Majumder, S., Mathur, A., & Javaid, A. Y., (2019). A study on recent applications of blockchain technology in vehicular ad-hoc network (VANET). In: *National Cyber Summit* (pp. 293–308). Springer, Cham.
14. Liu, J., Yu, Y., Jia, J., Wang, S., Fan, P., Wang, H., & Zhang, H., (2019). Lattice-based double-authentication-preventing ring signature for security and privacy in vehicular Ad-Hoc networks. *Tsinghua Science and Technology, 24*(16), 16716–1684.
15. Kumar, A., & Sinha, M., (2014). Overview on vehicular ad hoc network and its security issues. In: 2014 International conference on computing for sustainable global development (INDIACom) (pp.792–797). IEEE.
16. Lin, X., Lu, R., Zhang, C., Zhu, H., Ho, P. H., & Shen, X., (2008). Security in vehicular ad-hoc networks. *IEEE Communications Magazine, 46*(4), 88–916.
17. Samara, G., Al-Salihy, W. A., & Sures, R., (2010). Security analysis of vehicular ad-hoc networks (VANET). In: *2010 Second International Conference on Network Applications, Protocols and Services* (pp. 1616–1660). IEEE.
18. Sumra, I. A., Ahmad, I., & Hasbullah, H., (2011). Behavior of attacker and some new possible attacks in vehicular ad-hoc network (VANET). In: *2011 3rd International*

Congress on Ultra Modern Telecommunications and Control Systems and Workshops (ICUMT) (pp.1–8). IEEE.
19. Wasef, A., Lu, R., Lin, X., & Shen, X., (2010). Complementing public key infrastructure to secure vehicular ad-hoc networks [security and privacy in emerging wireless networks]. *IEEE Wireless Communications, 17*(16), 22–28.
20. Aggarwal, A., Shivani, G., & Mamta, M., (2021). A comparative investigation of consensus algorithms in collaboration with IoT and blockchain. *Transforming Cybersecurity Solutions using Blockchain* (pp. 115–140). Springer, Singapore.
21. Aggarwal, D., & Shivani, G., (2018). *A Comparative Study: Reviewing Performance of Routing Protocols in Mobile Ad-hoc Network*, 109–113.
22. Azam, F., Yadav, S. K., Priyadarshi, N., Padmanaban, S., & Bansal, R. C., (2021). A Comprehensive review of authentication schemes in vehicular ad-hoc network. *IEEE Access, 9,* 31309–31321.
23. Sharma, A., et al., (2020). A genetic improved quantum cryptography model to optimize network communication. *Recent Trends in Communication and Intelligent Systems* (pp. 47–54). Springer, Singapore.

PART V
Automation in Image Processing

CHAPTER 13

AUTOMATION OF IMAGE PROCESSING WITH ARTIFICIAL INTELLIGENCE AND PATTERN RECOGNITION METHODS

RAVI GORLI and M. V. KISHORE

Assistant Professor, Department of Information Technology, Anil Neerukonda Institute of Technology and Sciences, Visakhapatnam, Andhra Pradesh, India

ABSTRACT

True to myth, artificial intelligence (AI) and pattern recognition-based microscopy image processing technologies are gaining ground. Automation and dimensionality reduction are common. In the near future, I believe AI will have the greatest impact on collaborative microscopy. Many hours have been spent setting up equipment to ensure smooth procedures. Incorporating data from multiple sources allows you to expand your options. When AI is used in a specific application, it is only one of many possible solutions. Nonlinear mapping is rarely used because linear orthogonal multivariate statistical analysis is more common. Previously, most unsupervised automatic classification and HAC (hierarchical agglomerative clustering) research ignored partitioning (clustering). These are the most common image comparison and registration criteria. If you ask different people about the same issue, you may get different answers. Multiple solutions are redundant in this sense because they approach the problem differently. Expert systems, statistics, and neural networks can all be used to

The Fusion of Artificial Intelligence and Soft Computing Techniques for Cybersecurity.
M. A. Jabbar, Sanju Tiwari, Subhendu Kumar Pani, & Stephen Huang (Eds.)

solve supervised classification problems. A similar but slightly different strategy works better for specific problems. Consider all the clustering methods available, each of which operates in the parameter space while assuming certain cluster geometry assumptions. Although the second interpretation is frequently advanced, mine is preferred. End users must first examine the behavior of the tool before deciding which tool is best for the new application. The large number of options and the short deadline make this project difficult to complete. Computer vision and AI algorithms for microscopic imaging are relatively new. Only a few systematic comparative studies have been conducted to determine the most appropriate tool for a specific instance. The efficacy of a new tool, if it is perfect for your situation, can either capture or lose your attention if it turns out to be inappropriate. I anticipate more comparative studies as people become aware of and appreciate various techniques. When this occurs, AI and pattern recognition systems are reduced to primitive tools.

13.1 INTRODUCTION

In the field of microscope imaging, image processing and analysis are becoming increasingly crucial. The instruments used to accomplish this are derived from a wide range of fields and disciplines. For the most part, image analysis extensions of one-dimensional signal processing techniques are employed in this setting. When it comes to filtering operations, signal theory supplies the majority of the tactics employed. Filtering operations are procedures in which the frequency content of a picture is adjusted to achieve a specific goal. Therefore, image processing is largely linear when performed within this context. However, a great number of nonlinear tools have been proposed and are already in general usage, as opposed to linear tools. The application of mathematical morphology in image processing and image analysis is quite frequent, particularly when binary mathematical morphology and gray level mathematical morphology are used together. Even though these two schools of thought originated in distinct places, they have recently been brought together under the umbrella concept of picture algebra, which is rather intriguing.

The following piece will take a different approach than the last one. I am interested in delving deeper into the role of artificial intelligence (AI)-derived tools in the advancement of society. Digital image processing can be characterized as a whole through the application of AI to imaging. This is in contrast to the analysis of human brain images, which can be seen as characterizing human brain imaging with the help of AI. It is important to note that the AI methods I will discuss in this chapter are limited to images and can only be applied to photographs. This sets them apart from the previously mentioned approaches. Describing their unique nature without using the word "distinct" is challenging. The same applies to individuals who are more comfortable working with tools from the signal processing or mathematical morphological fields, and this holds true in both directions. Similarly, methodologies inspired by pattern recognition can also be described. This will be accomplished by investigating whether pattern recognition and AI approaches have found a place in the field of microscopic research, as well as in other fields. Therefore, it is worth considering whether future applications of these technologies will bring anything new to the field of microscope image processing, and whether previously unsolved problems will benefit from their use.

13.2 APPLICATIONS

In the introduction, it was stated that AI could be applied to any image analysis performed on a computer, which could be argued to be the case in this instance. For example, the number of applications that use one or more of the methodologies described in Part II, such as dimensionality reduction and learning data fusion, as well as uncertainty calculus, is likely to be limited. Dimensionality reduction and automatic classification are two such examples. Only a limited number of applications will be considered in the future. Classification is used in the majority of these applications in some form, whether it is pixel classification (segmentation), image classification, or structure classification within images, among other things. Other parts of these applications make use of pattern recognition techniques. Two excellent examples of pattern recognition applications in this type of environment are the recognition of simple geometric structures (using the Hough transform, for example) and the recognition of textural/fractal patterns (both using the Hough transform). Additional preliminary reports on the application of data fusion techniques will be presented at the conference.

13.3 PIXELS CLASSIFICATION

The ability to segment images is critical for image processing applications. According to the following example, a single image is divided into numerous pieces, such as foreground and background or different regions of a single item, which are then combined to restore the original image in its entirety as seen in the previous example. Additionally, segmentation has been demonstrated to be a persistent difficulty in a variety of contexts, in addition to being one of the most difficult tasks to perform. The optimal method to disassemble photographs into their basic pieces has been under investigation for nearly four decades by researchers. Imaging techniques like gray-level histogram calculation and gray-level global thresholding, which are both utilized in image segmentation from a single component, can be used to divide an image into many components.

There are several edge detection techniques, including maximum gradient edge detection, zero-crossing of the Laplacian edge detection, and others. Edge following is used to estimate the borders of objects or regions, a method of assessing homogeneity by enlarging the area under consideration.

A more complex approach is achieved by the use of deformable contour approaches, which combine the need for homogeneity with the requirement for discontinuity (called snakes). The watersheds technique, in particular, is used in the field of mathematical morphology, where it is particularly effective.

A multi-component image is becoming increasingly important in microanalysis, which is becoming more ubiquitous. The recording of several pictures (often referred to as maps) correlating to the presence of different chemical species in a material is possible using techniques other than X-ray and electron energy-loss microanalysis. These techniques include Auger and ion microanalysis, among others [1]. This indicates that the goal of segmentation would be the generation of a single labeled image, each piece of which would represent a different component of the specimen during the segmentation process.

Another subject in which multi-component pictures are useful is the field of electron energy-loss mapping, which is another use of multi-component pictures. In order to make reliable estimates of the true amplitude and location of the characteristic signals, it is necessary to capture a large number of photographs of the characteristic signals. This makes it possible to correctly complete the background model and subtraction operations. Among the most recent developments in this method are spectrum imaging, which involves taking a series of photographs (one photo for each energy channel in the spectrum) or spectra (one spectra for each energy channel in the spectrum) and stitching them together to form a single image of the spectrum (one per pixel in the image). The automated extraction of information from complex data sets, despite the fact that image segmentation is not always performed in an official capacity, data reduction and automatic classification approaches may potentially play a role in this context for the automated extraction of information from complex data sets.

When it comes to the analysis and segmentation of multiple-component photographs, the single-component image segmentation approach can be effectively used. Pixels are labeled using a feature vector. Each pixel is described by an array of D attributes arranged into an array of D dimensions, with each attribute organized into an array of D dimensions. As far as I'm aware, this has been the accepted way of conducting business in practice. The number of properties is the same as the number of signals that have been recorded. In this context, it is crucial to investigate the distinctions between unsupervised classification and supervised classification. Supervised classification can be used to separate classes if an expert is capable of guiding the system, for example, by supplying well-controlled examples that correspond to the numerous classes that must be separated, and the expert is capable of doing so. Whenever it is neither possible nor appropriate to establish a learning set, unsupervised categorization must be used in its place, as mentioned above.

13.3.1 EXAMPLE OF IMAGE SEGMENTATION

When it comes to segmenting multi-component images, interactive correlation partitioning (ICP) is the simplest and most successful method available (ICP, Section II.B.1). Using a two- or three-dimensional interactive cloud selection approach, the following are some examples of applications for this approach.

Several authors, including Paque et al. [2] have used scatterplots with three dimensions to illustrate their findings.

If you are looking for something more challenging, you can learn about the variations between different classes by using a training set selected by an expert. Later, the newly collected information is used to segment the remaining components of the multi-component image, which is a time-consuming procedure that takes several hours. Using the field of mineralogy as an illustration, Tovey et al. [3] provided an excellent example of how this approach can be successfully applied. Users-selectable training zones allowed students to gain an understanding of a range of minerals such as quartz, feldspar, and various types of feldspar and quartz. It was possible to construct a covariance matrix including statistical data on the gray level distributions of the various mineral classes by analyzing the different training areas. Any leftover pixels were categorized using a maximum likelihood classification approach, which was used to limit the amount of false positives in the final classification. To solve categorization difficulties, a post-processing step was performed on the labeled image before applying the quantification approaches (such as over-segmentation).

13.3.2 EXAMPLE OF UNSUPERVISED MULTI-COMPONENT IMAGE

The final segmentation result should be the same as in the previous example, but without the assistance of an expert. There will be clusters in the parameter space, but they will be separated based on pixel homogeneity. These clusters should be easily distinguishable through clustering. Scatterplots are useful when only two or three model components are required.

In multi-part assemblies, dimensionality reduction is performed first. In a high-dimensional environment, clustering becomes problematic when there are insufficient data points (i.e., when the number of components is large).

A study involved a spectroscopic X-ray fluorescence camera photograph (usually 50×50 pixels, 1,024 channels). The spectra were segmented to form pictures with usually 5 to 15 components. Then, dimensionality reduction was applied using principal component analysis (PCA). Analyzing the eigenvalues revealed three major categories that accounted for 89% of the variance. Further analysis of a granite sample identified microcline, albite, and opaque mineral groups, which accounted for 77% of the variance (*see* Figure 13.1).

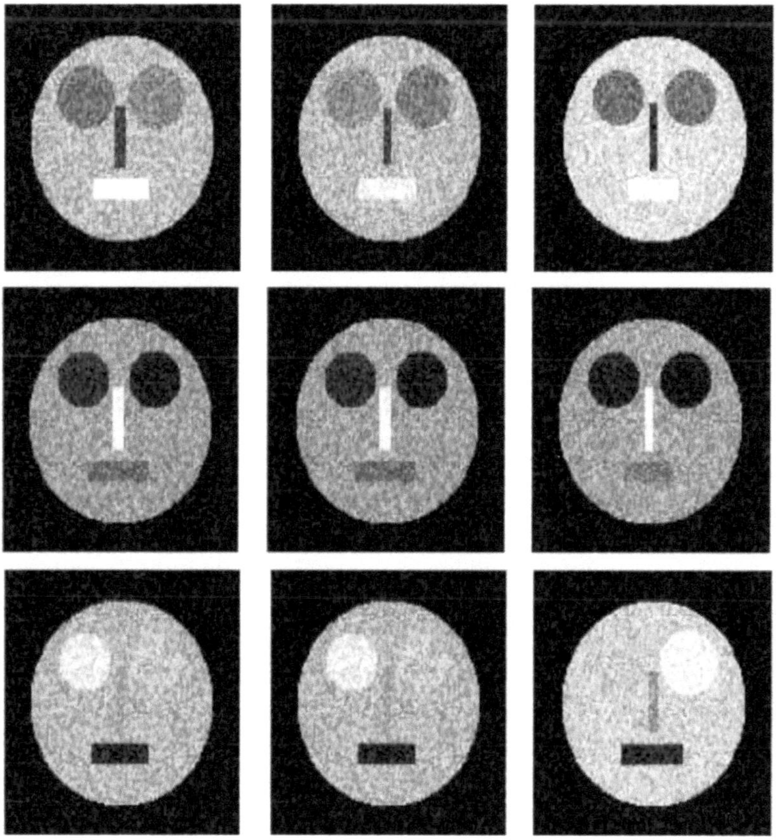

FIGURE 13.1 Classification in the context of macromolecule image classification.

Source: Reprinted from Quora.com The Science Space. Silk Road. https://www.quora.com/Why-do-IQ-tests-only-revolve-around-patterns

They classified photos using C-means clustering. By comparing signal intensities to object distances, they were able to recognize images and preprocess pixel data. They also utilized Roman glass studies to demonstrate the advantages of PCA + C-means clustering over hasty decision-making. Determining the ideal strategy requires further investigation, as making a global judgment seems impossible. To design the most effective technique for each application, data evaluation should be employed, as it is unlikely to be the same for all applications.

Additionally, Auger microanalysis was conducted. Haigh and colleagues developed Automatic Correlation Partitioning in 1997, which utilizes a D-dimensional intensity histogram for each material orientation to identify clusters. Statistical tests are then employed to validate the detection of histogram peaks.

Fluorescence microscopy can categorize pixels in an image. Similar datasets can be obtained using fluorescence microscopy, but with numerous images per fluorochrome rather than just one per sample. As in the preceding situation, a scatterplot and ICP are utilized. Other datasets include images with varying time and depth, or images with varying wavelength (spectrum images). Data from fluorescent video microscopy and confocal microscopy must be processed using modern data processing technology. Many datasets are multi-dimensional and must be reduced in order to be analyzed. This reduction was accomplished using PCA or CA. Large datasets can be reduced to Eigen images and scores using MSA. Datasets are not always orthogonal because the information sources that modify them are not always orthogonal. So, if deconstruction generates quantitative data, factor analysis (or oblique analysis) is required [4]. For this field, I choose Kahn and his team's project. The FAMIS approach can process time-lapse, multispectral, and depth-dependent confocal microscope images [5]. The researchers employed two separate fluorochromes on two different specimens to determine depth dispersion. They were also able to distinguish fluorochromes individually in dynamic series by observing the rate of photo bleaching. This technique aided chromosomal research in cytogenetics. They utilized fluorescence *in situ* hybridization to detect low copy numbers of HPV DNA. Additionally, they utilized three-dimensional (3D + time) confocal image sequences (FISH).

Instead of processing each pixel individually, Hannequin & Bonnet [6] employed multivariate statistical analysis to simultaneously handle multiple energy-filtered images. Based on their findings, Factorial filtering was used to remove non-essential components.

Noteworthy is Hannequin & Bonnet's [6] attempt to utilize MSA to provide quantitative results for the characteristic signal. Oblique analysis is necessary to establish a connection between one of the newly rotated axes and the chemical source.

MSA only receives backdrop images. These images are obtained by reconstructing and eliminating the background from the matched experimental shots. In another study, a picture taken at non-typical energy losses to estimate an unknown background. Each pixel carried a distinct signal as part of this process.

13.4 CLASSIFICATION OF IMAGES/SUB-IMAGES

It is necessary to consider the classification of images (or sub-images) when working with image sets as an alternative to pixel classification. This activity has a wide range of potential applications in both biology and materials science, among other fields. The first domain is concerned with the classification of 2D images of individual 3D macromolecules, which falls under its purview. In the second domain of classification, which also includes sub-units depicted as images in crystal images, biological or hard crystals are categorized as belonging to the second domain of classification.

13.4.1 MACROMOLECULES 2D VIEWS

Obtaining 3D structural information on macromolecular assemblies at a resolution adequate to distinguish between quaternary characteristics is one of the most challenging aspects in biological electron microscopy. In other words, the objective is to describe the particle architecture as precisely as X-ray crystallography (which is currently considered the gold standard) [7]. Consequently, the structure is statistically derived, with noise being eliminated through the vast number of comparable images.

A dataset containing hundreds or thousands of images can be heterogeneous due to the inclusion of distinct structures, multiple viewpoints of the same three-dimensional structure, or a combination of both factors. Automated classification is necessary to establish uniform classes of views based on particle type and viewing angle.

Images are high-dimensional objects; thus, it is advised to scale them down (a single image can be defined by as many attributes as there are pixels in it). This knowledge has been known to microscopists for about two decades. Scaling down removes redundancy while maintaining most of the critical information in the data, as pixels' intensity levels are closely related.

Nonlinear mappings will be covered in more detail later. For normalization and factorial image classification models. A data reduction factor of 1 to 400 reduces the spatial dimension tenfold (for 32×32 or 64×64 pixels). Uncorrelated noise accumulates into large components that the algorithm can easily identify and ignore. Additionally, CA can – (i) reject outliers; and (ii) minimize noise.

Studying molecular structure dynamics enables new research avenues, specifically trace structure analysis. The factorial space serves as the first step, where individual images now have fewer features, resembling scores in a smaller factorial area. Classification can be done without the assistance of a manager. Research has been conducted on C-means, DCC, and fuzzy C-means algorithms. Another study discovered that fuzzy techniques perform well for picture classification and established new criteria for judging the quality of a partition.

13.4.1.1 ALGORITHM

Their method included "a fast HAC algorithm backbone, a partition-enhancing post-processor, and some additional tweaks and interpretational assistance." The latter algorithm reduces the merger cost function when an element is shifted from one class to another. It is predicted that it will not be caught in within-class variance minima. Outliers will be removed from the data.

13.4.1.2 HYBRIDIZATION CLASS

The high computing demands of this method are a problem. Using HAC with C-means clustering can help reduce it. This results in numerous small classes (C'). Then the HAC merges the intermediate classes. Instead of utilizing the C-means approach, the dynamic clustering algorithm was proposed. It was then utilized again.

A range of innovative techniques for categorizing macromolecule images have recently been researched as alternatives to HAC and its modifications. To begin, I'll outline an effort to minimize dimensionality while still classifying data using self-organizing mapping (SOM). My next topic will be presenting the comparison and evaluation work we've recently done on various strategies (such as neural networks) for dimensionality reduction and classification. Maps that organize themselves are utilized.

Macromolecule pattern detection and classification were initially attempted by Marabini & Carazo [8], who also published their findings. Classification and alignment, two difficulties related to population variability, were to be solved in a single step. It turned out to be a significant success for them. To generate two-dimensional SOMs, only a modest number of neurons were required, and the final maps contained almost the same number of classes as expected. This was how they operated (i.e., 5×5 or 10×10). GroEL chaperonin particles were first aligned translationally but not rotationally in their initial application. Particles may be categorized using SOM's orientation in the plane, as shown by the researchers. The side views of TCP-1 complexes were the focus of their second use. Particle orientation categorization was shown to be successful even for less symmetrical particles, they found.

Using two sets of photographs, they constructed a third example that showed top views of the TCP-1 and TCP-1/actin binary complexes, respectively. They were able to sort such a wide range of information into 100 unique groups. The learning vector quantification (LVQ) method, created from SOM and applied to this dataset, was also provided in this chapter, but it was not stated. SOM was successfully applied by Barcena et al. [9] to study populations of SPP1 G40P helicase protein hexamers.

The fuzzy Kohonen clustering network (FKCN), which is a generalization of SOM for clustering applications based on fuzzy C-means ideas (also known as FLVQ for fuzzy learning vector quantization), was utilized to classify individual photos unsupervisedly. A comparison was conducted between the findings of FKCN and SOM in the rotating power spectrum, followed by interactive grouping into four groups: symmetry twice, symmetry three times, and the lack of it (Figure 13.1). FKCN exhibited less subjectivity than the other technique, and the agreement between the two methods was as high as 96%, according to their findings. The data collection, consisting of 388 photos with 3-fold and 6-fold symmetry, was re-examined to explore any new findings. SOM and FKCN were applied to the images themselves rather than the rotational power spectrum. The researchers observed that the FKCN method grouped the images into three classes, with the class corresponding to counterclockwise handedness being separated into two sub-classes, each holding a different level of three-fold symmetry. However, SOM failed to distinguish between the two.

In an attempt to apply SOMs for the study of electron micrographs of biological macromolecules, researchers were unsuccessful. Their network structure differs greatly from prior approaches because they used rings, double rings, and spheres instead of planes. This is where their strategy differs the most from others. An electron beam recognizes particle pictures in three dimensions, thus the researchers set out to achieve just that. It's crucial to note that they used the more sophisticated Fourier spectrum instead of photos collected in real time.

Dimensionality reduction and automatic categorization algorithms are studied and contrasted in this study. Macromolecule photo categorization has only studied a small percentage of the approaches available for dimensionality reduction and automatic classification. A wide number of approaches were assessed including the following:

- Multivariate PCA and Sammon's mapping are examples of dimensionality reduction techniques.
- Automatic categorization techniques include HAC, C-means, fuzzy C-means, ART-based neural networks, and the Parzen/watersheds method.

`In this context, a variety of specific themes and questions were addressed, including the ones listed below:

- Are nonlinear mapping methods more accurate and precise than linear mapping approaches?
- Is pre-processing the data with linear methods like PCA necessary when utilizing nonlinear approaches?

Realistic models (hypothetical structures with 47 and 48 subunits) and real-world pictures were used to address all of these concerns related to the GroEL chaperonin. The answers to these questions can be summarized as follows: When it comes to cloud reparability, PCA was found to be superior to nonlinear mapping when applied first. This superiority is attributed to PCA's ability to reject noise, as previously indicated. The purpose of PCA in this two-step technique is not to reduce the dimensionality of the data set to its smallest feasible value, but rather to reduce it to a dimension of approximately 10 and then proceed to further reduce it using nonlinear approaches. The results from Sammon's mapping and AANN were comparable, but they required a significant amount of processing time. It is undeniable that these algorithms need to be fine-tuned before they can be used to map thousands of items simultaneously in the real world. After multiple attempts, Guerrero et al. maintained the concept of using the entropy of a scatterplot that depicts inter-object distances (Dij) in the original space and the same distances after mapping (dij). There is a higher concentration of pairs on the main diagonal of the scatterplot when inter-object distances are preserved, while pairs (Dij, dij) are spread outside of the main diagonal (p(Dij, dij) is greater). Since the distances dij depend on the reduced space dimension (D'), the entropy is also affected by this. The derivative of entropy can therefore be determined as a function of the entropy constant D.' When this derivative achieves its maximum value, it appears to be a good signal of the optimal reduced space dimension. It is remarkable to learn that alternative clustering algorithms have been employed so infrequently in the context of macromolecule picture categorization. When Guerrero and colleagues ran simulations, a number of clusters were created, with HAC being the most extensively utilized technique in this context, coming out on the bottom. Not every data set would be the same, but the study's authors stated that it was specific to the data set they used. Because of this, they encourage users of automated systems to exercise caution. Before agreeing on a single categorization method and accompanying set of results, classification methods should be utilized with extreme caution and findings from numerous classification methodologies should be compared.

13.4.2 UNIT CELLS OF CRYSTALLINE SPECIMENS

13.4.2.1 BIOLOGY

Because 2D crystals are mathematically flawless structures, Fourier-based filtering methods are also based on this assumption. Other approaches are employed to address challenges at the unit cell level, as listed by Sherman et al. These methods include rigorous correlation averaging.

A categorization is necessary to exclude the possibility that some unit cells are not identical. The aforementioned procedures still apply. The research community indicates that MSA and HAC are becoming popular choices for therapy.

Recent applications in this field utilized patch averaging, SOMs, and multiscale analysis to investigate structural variation in two-dimensional crystals of the bacteriophage 29p10 connector. The objective of the study was to compare the accuracy of patch averaging followed by MSA analysis versus using SOM as an intermediate step between patch averaging and MSA analysis. In the 4×4 SOM, blank nodes between clusters divide the 16 neurons into three or four classes. This results in averaging patches from the same class. Using MSA on SOM code words instead of patches helps to understand the eigenvectors and their relationships.

Sherman & colleagues [10] discovered four distinct classes of gp32*I protein variety in a 4,300-unit cell crystal and averaged them independently (*see* Figure 13.2). Consider how each class is made up of unit cells. The differences between them aren't created by large-scale bending of the crystals, but rather by variances inside each unit cell. Many explanations have been proposed to explain class distinctions. Numerous hypotheses involve translation of the unit cell inside the crystal lattice, internal vibration of the unit cell molecule or molecules, and local tilting of the crystal plane. To improve the results of three-dimensional reconstruction, the authors anticipated to apply multiple averages (one for each class) rather than a single average. Their hypothesis was correct.

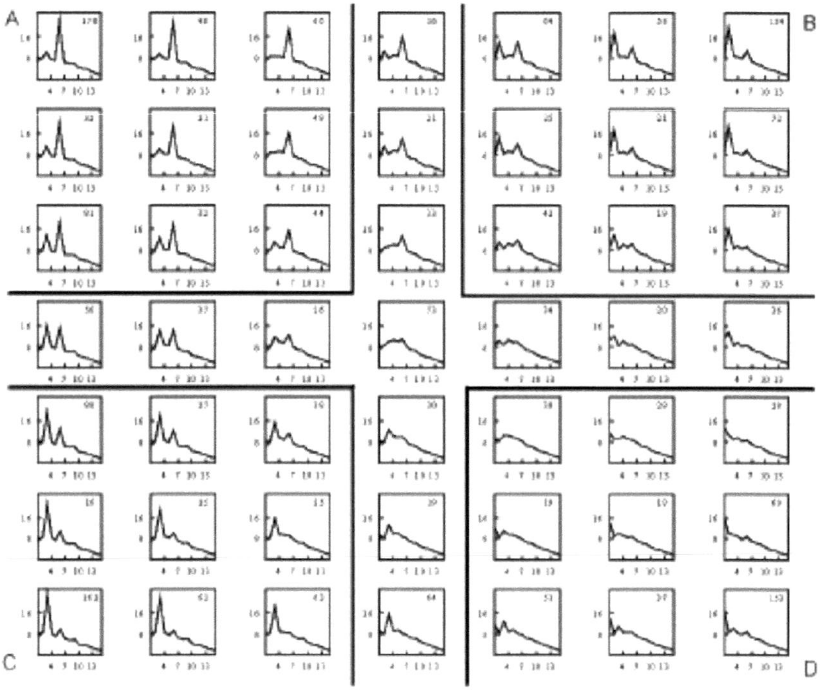

FIGURE 13.2 Illustration of automatic classification methods.

13.4.2.2 *MATERIAL SCIENCE*

The field of material science has significant applications in the study of crystal subunits, particularly in pattern recognition and classification.

High-resolution electron microscopy (HREM) allows for the comparison of images with subunits, which is particularly useful in medical imaging. This approach has been widely used to quantify changes in chemical composition on surfaces.

For this specific application, I will introduce a pattern recognition technique that was not covered in Part II but is closely related to the cross-correlation coefficient. Ourmazd and his team developed this approach [11].

In a multidimensional feature vector, the components are the gray levels of the pixels that make up a unit cell, for example, 30×30 = 900 pixels. The chemical composition of an averaged unit cell, located away from the interface, is compared to two reference feature vectors. This comparison reveals a plausible relationship between the unknown vector and the reference vectors, along with some reasonable indicators of that connection. Ourmazd and his colleagues use the following metric:

$$\chi = \frac{Arc\,\mathrm{Cos}\left(\theta_X\right)}{Arc\,\mathrm{Cos}\left(\Delta\theta\right)}$$

It is possible to estimate and plot the actual concentration corresponding to any unit cell of the interface if experimental conditions (such as defocus and specimen thickness) are properly adjusted. The Quantitem technique, a more advanced version of pattern recognition technology, was eventually developed [12, 13]. One method is to traverse an image using unit cells that define vectors. Variations in sample potential are explained by changes in sample thickness and/or chemical composition (in the parameter space). Individual unit cells along this path can track topographical changes or changes in the specimen's chemical composition after calibration. To obtain the composition function, perform the following operations – deconvolution of the difference image by the motive; least squares fit of the composition function, which is mathematically equivalent to "difference convolution."

This type of application in multivariate statistical analysis as part of a preliminary research project. We used simulated and experimental images of the GaAs/AlGaAs system, as well as linear multivariate statistical analysis, to determine the amount of Al in the interface. In contrast to techniques that rely solely on the calculation of a single indicator or a collection of indicators to predict the future, MSA produces intermediate outputs.

Aebersold et al. performed supervised classification on the interface of a Ni-based superalloy. They used PCA to reduce the parameter space's dimensionality to learn class centers and extensions. They selected representative zones for each class. The researchers next tried three approaches for classifying each cell in a unit.

Information can be classified using MDCM, ML, or PE probability. Our PCA and ML studies revealed that the findings were unaffected by whether or not we used a larger number of components (D').

13.5 OBJECTS DEDUCTED IMAGES

Aside from the two extreme situations, another use for classification systems is the categorization of objects in pictures after segmentation (pixel classification and image classification). Shape, texture, color, and other characteristics can all be used to recognize an object. Applications like particle and defaults analysis fall into this area. It includes the computation of features, their reduction and selection, and classification of supervised and unsupervised data. Because addressing the issue of feature computation would require a book, I will not go into great length here. I'll only give a few instances here.

For the first time, AI applied to solve the twin identification problem. Kohlus & Bottlinger [14] used neural networks to categorize particles based on their Fourier coefficients (multi-layer feedforward NN and SOMs). Nestares et al. used texture segmentation to automate the process of segmenting areas subjected to brief laser pulses in Sb materials, with encouraging results. The ISODATA version of the C-means algorithm described in part II was used to group pixels of similar size and shape. Livens et al. [15] used texture analysis to characterize corrosion pictures, as described below. They define their features using wavelet decomposition and then rotate them in various ways. A supervisor oversaw the organization's coding system. It uses learn vector quantization (LVQ). This method frequently outperformed other methods like the Gaussian quadratic classifier and the K-nearest neighbors (KNNs) classifier.

Neural networks and expert systems are used to study microscopic wear particles in rubber. Shape and texture were used to construct the features. Dividing surfaces into smooth, rough, striated, and pitted. A computer vision system and a knowledge-based system are being combined to help produce an integrated system that can predict a machine's breakdown based on its past. Texture analysis and classification have long been utilized in biology, such as hematology [16]. They can also be used to study chromatin architecture for cancer prognosis. For cervical cytology cells, Smeulders et al. [17] used nuclear chromatin pattern textural qualities to characterize them.

The entropy of co-occurrence matrices compares to examine feature capabilities in supervised learning. The jackknife and leave-one-out discriminant analysis methods were used to select the top four attributes. These four parameters achieved a 90% success rate in classifying tumor specimens and showed minimal changes in the tumor nucleus. Using confocal microscopy, three-dimensional images are obtained.

13.5.1 HOUGH TRANSFORMATION

This technique was not applied to electron microscopy until 1989, when researchers were able to study electron diffraction patterns and specifically detect low-contrast Kikuchi lines for the first time. This method has the potential to be used for investigating electron backscattering patterns in the future.

Based on these observations, it is concluded that the automated Hough transform was capable of detecting bands with an accuracy comparable to that of human operators.

Recently, there has been increased interest in this technique for the automatic interpretation of convergent beam electron diffraction patterns (CBED), which were created and first published in 1999. Figure 13.3 illustrates a zone axis CBED device made of aluminum, demonstrating the process in action.

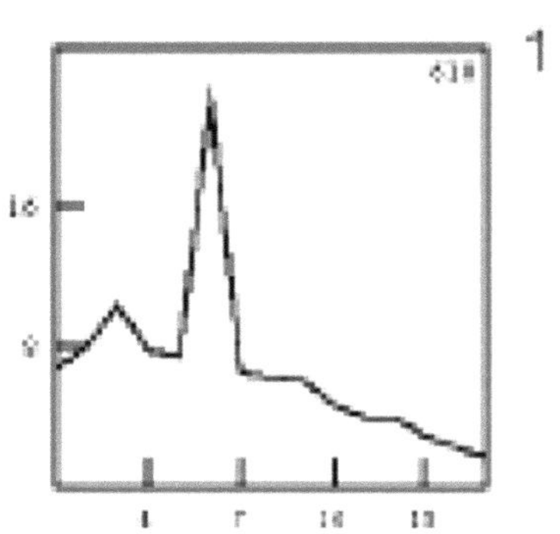

FIGURE 13.3 Illustration of the use of the Hough transform.

13.5.2 FRACTAL ANALYSIS

Several studies utilized digital annular dark field STM (scanning transmission electron microscopy) to observe aggregated iron grains (STEM). Another study employed SEM to classify airborne particles into eight different categories. The shape and chemical composition of the particles were determined using the geometrical moment invariants and X-ray spectra. When used for particle grouping, this neural network outperforms the hierarchical ascendant classification method.

Individual aerosol particles (fly ash particles) imaged by SEM were found to be categorizable using the fractal dimension. SEM backscattered electron imaging was utilized to determine the shape of microparticles (silver chloride crystals with 'T-grains' and aerosol particles) visible in Fourier analysis and fractal analysis. The samples were also subjected to energy-dispersive X-ray microanalysis and shape analysis. Additionally, a similar approach was employed by Oleshko et al. [18] for the analysis of quasi-fractal many-particle systems, including colloidal Ag particles, along with the detection of percolation networks of Ag filaments by TEM.

Other examples of the application of the fractal concept can be found in the following sources: Another study discusses fractal development mechanisms of soot and fractal patterns in annealed sandwich Au/a-Ge/Au, which are references to the author's work in this area. Another study is conducted corrosion crack stress multifractal analysis. Another study is used scanning tunneling microscopy to study surface fractal character and calculated microstructural parameters using random-space microstructural parameters. Herman & Ohser [19] published their findings in 1993, and Herman et al. [20] published their findings in 1994. Take the following into consideration when studying biology.

In this context, biological fractals are used. The fractal dimension of the cell outlines was utilized to segment the image in Rigaut's (1988) study on Image Segmentation Based on Fractal Geometry and Mathematical Morphology.

An analysis of microscopically recorded photos using fractal geometry. A comparison of human T-lymphocytes and hairy leukemic cells is being conducted to see how similar they are to one another and how they differ [22]. Researchers in the '90s found that characterization of the chromatin texture in the turtle retina [23]; analysis of the amacrine, ganglion, horizontal, and bipolar cell complexity, and scaling characteristics in the turtle retina.

13.5.3 IMAGE COMPARISON

For iterative structural refinement, high-resolution electron microscopy can be used to compare simulated and experimental images. Although many electron microscopists value this aspect highly, I won't go into detail about it here because image comparison methods rely almost entirely on the least squares criterion. Yet another topic for me to write about.

In another study, cross-correlation techniques can be utilized on grain boundary structures photographed by HREM to determine similarity between shots, and the coefficient of correlation can be used to do so. Using image cross-correlations, we were able to determine which positions inside the grain boundary and bulk material were similar, as well as how similar they were on a scale of 1 to 10. Despite the fact that this pattern identification technique has been demonstrated to be advantageous, the authors noted various drawbacks, such as the fact that a change in the correlation coefficient does not necessarily indicate what type of structural deviation has taken place. A parametric explanation for the distortion was found to be superior by the study's authors than earlier findings.

Alternatives to the least squares criterion technique, although uncertain, will be helpful in image comparison and analysis.

13.6 DATA FUSION

To the best of my knowledge, there has been no significant utilization of data fusion in microscope imaging, which refers to the merging of varying degrees of belief presented by multiple images to generate a conclusion, or the integration of mathematically diverse degrees of belief provided by several images to reach a judgment. Researchers often rely on empirical methods to address specific challenges.

Creating a focused single image with high magnification light microscopy is a challenging task. For instance, Farkas et al. [25]; and multi-modality techniques is explored in light microscopy. The use of fluorescence and immunofluorescence microscopy on the same specimen allows for complementary data collection. Bright field (BF), phase contrast (PC), and differential interference contrast (DIC) microscopy are among the available methods. Glasbey and Martin demonstrated approaches for pre-processing data sets using triple modality images (BF, PC, and DIC).

These technologies primarily focused on image alignment, as changes in optical settings could cause image shifts. The authors also attempted to determine if the content of the photographs exhibited correlation or anti-correlation. Color images with only one component in each red, green, and blue (RGB) channel were employed to aid in distinguishing between images. However, quantitative analysis remains considerably challenging. The authors suggested using Principal Components Analysis (PCA) for this project. As a result of their research, they discovered that the correlation of the BF DIC and PC images, and the inverse correlation, accounted for almost 74% of the entire variation in their case study. This component (which accounts for 23% of variation) shows a correlation between the three photographs that was difficult to notice visually. The residual component can be used to generate an optical thickness image using the anti-correlated part (explaining only 3% of the variance).

Another study used confocal microscopy to create images that combined fluorescence and transmitted light. Three-dimensional scatter-plots and thresholding were developed to cope with multimodal data and detect pixel clustering. Despite being a first draft, it gives a good picture of where things are going. Combining light microscopy and fluorescence pictures can improve light microscopy results.

There are several methods for combining data from different sources in electron microscopy. This work uses STEM spectroscopy, which includes dark field imaging, Z-contrast imaging, and electron energy loss spectroscopy [26].

Combining X-ray and EEL microanalysis [27–29], in addition to surface analysis equipment with several sensors, transmission electron microscopy, or a combination of X-ray analysis and microscopy [1, 3]. Until now, the user's brain has been used to combine data rather than computer data fusion technologies. The following papers provide more in-depth examples of clinical research.

Barkshire et al. [30] used image correlation to adjust for fluctuations in beam current in quantitative surface microscopy (multispectral Auger microscopy). Another study claim that four backscattered electron detectors combined to establish topographical contrast.

Leapman et al. [31] used data fusion to tackle the low signal-to-background problem in EELS. Because cryosectioned cells had a low calcium content, the researchers set out to measure it. It was difficult to average multiple spectra of the recorded spectrum-image at the moment of recording. They required help figuring out which pixels were homogeneous (and thus could be averaged securely). Simultaneously recording nitrogen map-able signals enabled them to discriminate between distinct regions of interest (ER, mitochondria) based on nitrogen concentration fluctuations. The average spectra allowed them to calculate the average calcium content in each of the target locations.

13.7 CONCLUSION

Artificial intelligence (AI) and pattern recognition-based technologies appear to be gaining favor for processing microscope pictures, contrary to conventional opinion. Automatic classification and dimensionality reduction, for example, are frequently utilized. I believe collaborative microscopy will benefit the most from AI in the near future. The majority of the effort has gone into setting up the necessary instruments for these procedures to function. The next stage is to expand your options for combining data from various sources, which is where data fusion techniques come in handy. When AI techniques are currently in use in an application, only one of several feasible solutions is usually investigated. Nonlinear mapping is rarely employed since linear orthogonal multivariate statistical analysis is more popular. Partitioning has been mostly overlooked in unsupervised automatic classification, similar to hierarchical ascending classification (clustering). While other picture comparison and registration criteria exist, the least squares criterion and associated correlation coefficient are the most extensively employed. A variety of solutions to a single problem can be perceived in different ways. The multiple solutions are redundant because each one addresses the problem in a unique manner. Expert systems, statistics, and neural networks may all approach supervised classification problems similarly. However, each method is better suited for solving a specific type of problem compared to a similar but slightly different strategy. Consider the various clustering methods, all of which make implicit assumptions about the cluster geometry in the parameter space. Despite some repetition, I generally prefer the second interpretation. Potential end users must study the behavior of the tool to determine which one best suit the needs of their new application. Given the large number of options and the urgency of the search, this would be a challenging task. AI/pattern recognition algorithms for microscope pictures are still in their early stages in this regard. Despite the abundance of tools discussed in previous sections of this study, few systematic comparative studies have been conducted to evaluate the optimal tool for a given case. A new tool can either spark interest (if it is ideal for the situation) or leave one disappointed and rejected (if it is not). I anticipate more comparative studies as more people become familiar with and understand the various approaches. When this happens, AI and pattern recognition systems will become basic tools with limited capabilities.

KEYWORDS

- **algorithm**
- **artificial intelligence**
- **hierarchical agglomerative clustering**
- **Hough transformation**
- **image processing**
- **image segmentation**
- **macromolecules**
- **pixels**

REFERENCES

1. Le Furgey, A., Davilla, S., Kopf, D., Sommer, J., & Ingram, P., (1992). *J. Microsc., 165*, 191–223.
2. Paque, J. M., Browning, R., King, P. L., & Pianetta, P., (1990). In: Michael, J. R., & Ingram, P., (eds.), *Microbeam Analysis* (pp. 195–198). San Francisco Press, San Francisco.
3. Tovey, N. K., Dent, D. L., Corbett, W. M., & Krinsley, D. H., (1992). Scanning *Microsc. Suppl., 6*, 269–282.
4. Malinowski, E., & Howery, D., (1980). Factor Analysis in Chemistry. Wiley-Interscience, New York.
5. Di Paola, R., Bazin, J. P., Aubry, F., Aurengo, A., Cavailloles, F., Herry, J. Y., & Kahn, E., (1982). *IEEE Trans. Nucl. Sci. NS, 29*, 1310–1321.
6. Hannequin, P., & Bonnet, N., (1988). *Optik, 81*, 6–11.
7. Harauz, G., (1988). In: Gelsema, E. S., & Kanal, L. N., (eds.), *Pattern Recognition in Practice* (pp. 437–447). North-Holland.
8. Marabini, R., & Carazo, J. M., (1994). *Biophysical Journal, 66*, 1804–1814.
9. Barcena, M., San, M. C., Weise, F., Ayora, S., Alonso, J. C., & Carazo, J. M., (1998). *J. Mol. Biol., 283*, 809–819.

10. Sherman, M. B., Soejima, T., Chiu, W., & Van, H. M., (1998). *Ultramicroscopy, 74*, 179–199.
11. Ourmazd, A., Baumann, F. H., Bode, M., & Kim, Y., (1990). *Ultramicroscopy, 34*, 237–255.
12. Kisielowski, C., Schwander, P., Baumann, P., Seibt, M., Kim, Y., & Ourmazd, A., (1995). *Ultramicroscopy, 58*, 131–155.
13. Maurice, J. L., Schwander, P., Baumann, F. H., & Ourmazd, A., (1997). *Ultramicroscopy, 68*, 149–161.
14. Kohlus, R., & Bottlinger, M., (1993). *Part. Part. Syst. Charact., 10*, 275–278.
15. Livens, S., Scheunders, P., Van De, W. G., Van, D. D., Smets, H., Winkelmans, J., & Bogaerts, W., (1996). *Microsc. Microanal. Microstruct., 7*, 1–10.
16. Landeweerd, G. H., & Gelsema, E. S., (1978). *Patt. Rec., 10*, 57–61.
17. Smeulders, A. W., Leyte-Veldstra, L., Ploem, J. S., & Cornelisse, C. J., (1979). *J. Histochem. Cytochem., 27*, 199–203.
18. Oleshko, V., Kindratenko, V. V., Gijbels, R. H., Van, E. P. J., & Jacob, W. A., (1996). *Mikrochim. Acta Suppl., 13*, 443–451.
19. Hermann, H., & Ohser, J., (1993). *J. Microsc., 170*, 87–93.
20. Hermann, H., Bertram, M., Wiedenmann, A., & Herrmann, M., (1994). *Acta Stereol., 13*, 311–316.
21. Keough, K. M., Hyam, P., Pink, D. A., & Quinn, B., (1991). *J. Microsc.,163*, 95–99.
22. Nonnenmacher, T. F., Baumann, G., Barth, A., & Losa, G. A., (1994). *Int. J. Biomed. Comput., 37*, 131–138.
23. Fernandez, E., Eldred, W. D., Ammermüller, J., Block, A., VonBloh, W., & Kolb, H., (1994). *J. Compar. Neurol., 347*, 397–408.
24. Paciornik, S., Kilaas, R., Turner, J., & Dahmen, U., (1996). *Ultramicroscopy, 62*, 15–27.
25. Farkas, D. L., Baxter, G., BeBiaso, R. L., Gough, A., Nederlof, M. A., Pane, D., Pane, J., et al., (1993). *Annu. Rev. Physiol., 55,* 785–817.
26. Engel, A., & Reichelt, R., (1984). *J. Ultrastruct. Res., 88*, 105–120.
27. Leapman, R., Fiori, C., Gorlen, K., Gibson, C., & Swyt, C., (1984). *Ultramicroscopy, 12*, 281.
28. Oleshko, V., Gijbels, R., Jacob, & Alfimov, M., (1994). *Microbeam Analysis, 3*, 1–29.
29. Oleshko, V., Gijbels, R., Jacob, W., Lakière, F., Van, D. A., Silaev, E., & Kaplun, L., (1995). *Microsc. Microanal. Microstruct., 6*, 79–88.
30. Barkshire, I. R., El-Gomati, M. M., Greenwood, J. C., Kenny, P. G., Prutton, M., & Roberts, R. H., (1991). *Surface Interface Analysis, 17*, 203–212.
31. Leapman, R. D., Hunt, J. A., Buchanan, R. A., & Andrews, S. B., (1993). *Ultramicroscopy, 49*, 225–234.
32. Arndt-Jovin, D. J., & Jovin, T. M., (1990). *Cytometry, 11*, 80–93.
33. Bandarkar, S. M., Koh, J., & Suk, M., (1997). *Neurocomputing, 14*, 241–272.
34. Bright, D. S., Newbury, D. E., & Marinenko, R. B., (1988). In: Newbury, D. E., (ed.), *Microbeam Analysis* (pp. 18–24).
35. Buchanan, B. G., & Shortliffe, E. H., (1985). *Rule-Based Expert Systems*. Addison-Wesley, Reading.
36. Dubois, D., & Prade, H., (1985). *Theorie des Possibilités. Application à la Représentation des Connaissances en Informatique*. Masson, Paris.

37. Duda, R. O., & Hart, P. E., (1973). *Pattern Classification and Scene Analysis*. Wiley Interscience, New York.
38. Guersho, A., (1979). *IEEE Trans. Info. Proc., 25*, 373–380.
39. Haigh, S., Kenny, P. G., Roberts, R. H., Barkshire, I. R., Prutton, M., Skinner, D. K., Pearson, P., & Stribley, K., (1997). *Surface Interface Analysis, 25*, 335–340.
40. Han, J. H., Koczy, L. T., & Poston, T., (1994). *Patt. Rec. Lett., 15*, 649–658.
41. Haralick, R. M., (1979). *Proc. IEEE, 67*, 786–804.
42. Harauz, G., Chiu, D. K. Y., MacAulay, C., & Palcic, B., (1994). *Anal. Cell Pathol., 6*, 37–50.
43. Hawkes, P. W., (1993). *Optik, 93*, 149–154.
44. Hawkes, P. W., (1995). *Microsc. Microanal. Microstruct. 6*, 159–177.
45. Heindl, E., Rau, W. D., & Lichte, H., (1996). *Ultramicroscopy, 64*, 87–97.
46. Henderson, R., Baldwin, J. M., Downing, K. H., Lepault, J., & Zemlin, F., (1986). *Ultramicroscopy, 19*, 147–178.
47. Henderson, R., Baldwin, J. M., Ceska, T. A., Zemlin, F., Beckmann, E., & Downing, K. H., (1990). *J. Mol. Biol., 213*, 899–929.
48. Herbin, M., Bonnet, N., & Vautrot, P., (1996). *Patt. Rec. Lett., 17*, 1141–1150.
49. Hillebrand, R., Wang, P. P., & Gösele, U., (1996). *Information Sciences, 93*, 321–338.
50. Hough, P. V. C., (1962). U.S. Patent 3,069,654
51. Huang, S. H., & Endsley, M. R., (1997). *IEEE Trans. Syst. Man Cybern., 27*, 465–474.
52. Hÿtch, M. J., & Stobbs, W. M., (1994). *Microsc. Microanal. Microstruct., 5*, 133–151.
53. Illingworth, J., & Kittler, J., (1988). *Comp. Vision Graph. Im. Proc., 44*, 87–116.
54. Kenny, P. G., Barkshire, I. R., & Prutton, M., (1994). *Ultramicroscopy, 56*, 289–301.
55. Kohonen, T., (1989). *Self-Organization and Associative Memory*. Springer, Berlin.
56. Kraaijveld, M. A., Mao, J., & Jain, A. K., (1995). *IEEE Trans. Neural. Net., 6*, 548–559.
57. Kramer, M. A., (1991). *AIChE Journal, 37*, 233–243.
58. Kullback, S., (1978). *Information Theory and Statistics*. Smith, Gloucester, MA.
59. Lebart, L., Morineau, A., & Warwick, K. M., (1984). *Multivariate Descriptive Statistical Analysis.* Wiley, New York.
60. Lippmann, R., (1987). *IEEE ASSP Magazine April 1977*, 4–22.
61. Mandelbrot, B. B., (1982). *The Fractal Geometry of Nature*. Freeman, San Francisco, CA.
62. Marabini, R., & Carazo, J. M., (1996). *Patt. Rec. Lett., 17*, 959–967.
63. Mitra, S., & Pal, S. K., (1996). *IEEE Trans. Syst. Man Cybern., 26*, 1–13.
64. Nestares, O., Navarro, R., Portilla, J., & Tabernaro, A., (1996). *Ultramicroscopy.*
65. Pal, N. R., Bezdek, J. C., & Tsao, E., (1993). *IEEE Trans. Neu. Net., 4*, 549–557.
66. Shen, D., & Ip, H. H. S., (1999). *Patt. Rec., 32*, 151–165.
67. Shepard, R. N., (1966). *J. Math. Psychol., 3*, 287–300.
68. Tence, M., Chevalier, J. P., & Jullien, R., (1986). *J. Physique, 47*, 1989–1998.
69. Tickle, A. B., Andrews, R., Golea, M., & Diederich, J., (1998). *IEEE Trans. Neu. Nets, 9*, 1057–1068.
70. Hammel, M., & Kohl, H., (1996). *Inst. Phys. Conf. Ser., 93*, 209–210.
71. Hoekstra, A., & Duin, R. P., (1997). *Patt. Rec. Lett., 18*, 1293–1300.
72. Aldi, & Hornik, K., (1989). *Neural Networks, 2*, 53–58.

INDEX

Q

R

S